SPACE

THE FRONTIERS OF MODERN DEFENCE

SPACE

THE FRONTIERS OF MODERN DEFENCE

Squadron Leader **K K Nair**

Introduction by
Air Commodore **Jasjit Singh** AVSM, VrC, VSM, IAF (Retd)

IOWV
Knowledge World

in association with

Centre for Air Power Studies

The **Centre for Air Power Studies** is an independent, non-profit, academic research institution established in 2002 under a registered trust to undertake and promote policy-related research, study and discussion on the trends and developments in air power and the aerospace arena for civil and military purposes. Its publications seek to expand and deepen the understanding of air power and aerospace issues without necessarily reflecting the views of any institution or individuals except those of the author.

Jasjit Singh
Director
Centre for Air Power Studies
P-284, Arjan Path
Subroto Park
New Delhi-110010

Tele: (91-11) 25699131
E-mail: airpowerindia@yahoo.co.in

Knowledge World
5A/4A Ansari Road, Daryaganj, New Delhi 110002

Published in India by Knowledge World

ISBN: 81-87966-44-0

Typeset and Printed by Hardev Enterprises, Nirankari Colony, Delhi-110009. Published by Kalpana Shukla, Knowledge World, 5A/4A Ansari Road, Daryaganj, New Delhi-110002.

Contents

Introduction

Today, the ultimate high ground is space.
— General Joseph W. Ashy, USAF

Importance of space in national techno-economic, social-commercial and military-strategic life has been rapidly increasing during the past two decades. India itself has a robust space programme which is essentially geared toward scientific and developmental goals. This in itself increases the vulnerability for our country to any hostile action that might seek to damage, degrade or deny space capabilities so painstakingly built up over the decades at great cost against great odds. India's dependence on space for vital economic purposes has been growing rapidly during the past decade or so. Any serious damage or degradation of space resources for commercial and peaceful uses would have a major negative impact on India's security and well-being.

Space capabilities, therefore, are becoming critically essential for national development, economic well-being, commerce, and everyday life, besides assuming a crucial component of successful military operations. In fact space is emerging as a military and economic centre of gravity for our information-dependent, society, business and military forces. In short, life on earth is becoming inextricably linked to space. Global trends indicate that space is increasingly becoming an economic centre of gravity, the loss of degradation of which would cripple commerce, finance and numerous other private and public activities. The extent and expanse of our own expanding space capabilities clearly point to the concurrent vulnerabilities that are intrinsic to such acquisitions. The scale of the worldwide value of commercial space, which was estimated to be $11.5 billion in 1991, had increased to $57 billion by 2001. Assets of such high values obviously attract potential threats of many dimensions. The lesson of history, on the other hand, is clear that wherever serious threats to national economic interests arise, military force would be necessary to protect them in the best manner possible. This generates the rationale for military utilisation (and potential militarisation/weaponisation) of space, besides the intrinsic characteristics of space environment and resources to enhance terrestrial military capabilities and national defence.

Military, civil, and commercial space sectors are converging. Military and commercial uses of space will become increasingly vital national interest for India. Space is the 21st century's high ground impacting as it does on everyday lives of human beings and playing a crucial role in human development activities. Its role in national defence and security is equally crucial although many would say that this has not been adequately appreciated in India. As we move toward greater human development, we must ensure that utilisation of space for economic and development purposes is maximised. At the same time it is necessary to make use of existing space resources and create new ones for national defence and security. The bipartisan Standing Committee on Defence of the parliament has been emphasising this point for a numbers of years.

Navies and armies have evolved to protect national interests and investments. Industrialisation of war since the 18th century led to expansion of military power to acquire territories, resources and commerce, in turn fuelling the industrialisation of war even further. Armies and navies expanded combat power on the strength of newer military technologies and concomitant tactics and strategy. In turn these were used to acquire and protect the colonies, commerce and economic interests of those who could afford it, and who had the strategic vision to pursue geo-economics through geo-strategies and build military-commerce synergy for competitive advantages.

Early in the 20th century, air power emerged to strike and protect the surface forces that had traditionally protected national interests and investments. But very soon its role expanded to that of a separate instrument of warfare, protecting national interests and ensuring the freedom of action for air and surface forces. If air power transformed military capabilities and warfare in the 20th century, space power holds the promise to do more in the 21st century. Given the nature and extent of development of capabilities in space by various countries, it is obvious that besides the critical importance of space for national economic activities, **space power has become a precondition to control the land, sea and/or air power besides providing the foundations of nuclear deterrence.**

Although space power has supported land, sea and air operations in its early decades during 1970s and 1980s, its primary function had been perceived in terms of nuclear deterrence and possible war. It was the Gulf War 1991 that started to make a major difference to the role of space in conventional warfare in the air and on land/sea. This war was a watershed even in military space applications because for the first time, space systems were both integrated to the conflict and critical to the outcome of the war. While space resources play a crucial role in sustaining credible nuclear deterrence, the same resources have immense potential and capabilities to prosecute a "high-tech" conventional war with tremendous competitive advantage to the side that is able to employ them effectively.

As we move forward in the 21st century it is inevitable that space will become another medium of warfare besides assuming the important role of protecting the country's commercial assets in this medium. Developments in space capabilities and resources even in Asia are clear evidence that this phenomenon is no longer limited to a handful of developed industrialised countries. The United States would undoubtedly remain the leading space power for many decades to come. It can be nobody's argument that we need to acquire capabilities similar to that of the United States, or for that matter, other industrial states. But if China is seen as a point of reference closer home, one finds that it is rapidly emerging as a space power and its capabilities may also be placed at the disposal of other countries for strategic and/or commercial reasons. China, of course, has adopted a space vision "to become a world leader in the field of space science and exploration of outer space" within two decades. Its 10-year space objectives spelt out in 2000 include integrated military-civil earth observation systems, independently operated indigenously built satellite broadcasting systems, and independent satellite navigation and positioning systems besides new generations of satellites and launch vehicles. Its 2004 Defence White Paper clearly spelt out the goal of its military policy to "win local wars under conditions of informalisation". And that informationalisation almost entirely depends on exploiting space assets and capabilities.

Space has been militarised for 40 years used for transit and even dispensing multiple warhead of the most devastating weapons of mass destruction carried on ballistic missiles that could fly at even intercontinental ranges. (But this should not be confused with "weaponisation" of space which is not permissible according to current laws.) Reconnaissance, surveillance, warning, communications, weather, and most recently, navigation satellites were designed and deployed a serve national security needs. Systems related to national security have dominated space, but this dominance is eroding rapidly. The increasing number of countries and consortia turning to space to provide and receive services, and to generate wealth, will force nations to adapt to this emerging environment. Achieving advantage in space power before and during conflicts will be critical to success on the battlefield. Traditional military missions for land, sea, and air (like communication, surveillance and war fighting in air, land and/or at sea, etc.) are migrating to space.

It is against this context that we in India need to address the role of space in our national defence and security. The Ministry of Defence had confirmed to the Standing Committee on Defence (1999-2000) of Thirteenth Lok Sabha that "It has been proposed to impart a defence orientation to our successful space programme by including surveillance sensors, communication and navigation satellites". Two issues require attention. Firstly, it needs to be recognised that, as the US doctrine states, "Air and space constitute a seamless medium and that space capabilities should be fully integrated into air power". This is no doubt the reason why the Standing Committee on Defence of the parliament in successive reports have emphasised the importance of establishing an Aerospace Command in the Indian Air Force. Secondly, it is necessary to develop a sound doctrine and strategy for development of space capabilities and their utilisation for defence. This should help to develop space forces which may be defined as the space and terrestrial systems, equipment, facilities, organisations, and human resources necessary to access, use, and if directed, control space for national security. These tasks require suitable organisation to manage the development and employment of space capabilities for the successful effective employment of air power.

Squadron Leader KK Nair has carried out an excellent job of handling a complex and difficult subject with care and objectively. Nair as a young serving officer of the Indian Air Force who has been on study leave with the Centre for Air Power Studies for nearly two years as a Research Fellow. The views and conclusions expressed in the book are his own and do not necessarily represent those of any organisation or institution. It is not only Nair's first book, but this is also the first book in India by an

Indian author which would be of great value to the professionals as well as lay readers interested in such issues. It remains my pleasant duty to thank Air Headquarters for their generous support in a variety of ways to assist the Centre and making Nair available for what is undoubtedly a very valuable pioneering study.

15th March, 2006
New Delhi

Jasjit·Singh
Air Commodore (retd)
Director
Centre for Air Power Studies

Chapter 1
Evolution of Space for Military Uses

Outer space has always fascinated human beings. Numerous ancient myths and legends of various cultures recount fantastic and at times visionary accounts of the realm of space being used for conflict resolution apart from divine observation, intervention and punishment. The gods above perennially observed and monitored terrestrial activities, intervening at will to deliver punishment or reward as the case may be. Space-based military force application in terms of celestial reprimand from the heavens is a common recurrent theme in most ancient religious and mythical literature. Hence it is not surprising that ever since the realm above the atmosphere opened up, man has been devising ways and means of utilising it for military progress followed by economic, scientific and social advancement. In fact, it was the pursuit of better military capabilities like observation, surveillance as well as the improved efficiency of hurling ordnance on to the enemy afforded by the 'high ground' that led humankind to go higher and higher, beyond horses and elephants to air-breathing machines and then into the ultimate realm of space. Ever since the dawn of civilisation in terms of military conflict till the present, military doctrine places extraordinary emphasis on acquisition of high ground for successful prosecution of military operations. The difference between victory and defeat often relies on how effectively a military commander can utilise the advantage of height and that explains the perennial quest for high ground.

Historical Evolution

Early History

Thus from an early age, while the military uses of space were not formalised in any form, as an extension of the high ground the realm of space beyond atmosphere was broadly understood to be useful for delivery of military ordnance and observation (perspective) of terrestrial activities. As a related corollary, ever since the worth of space-based military applications became more apparent, realisation dawned that it was a realm worth fighting and establishing control over in order to wield decisive power over the enemy. Thus most of the early space-related military developments explored missions in areas concerned with military force application in terms of delivery of munitions as well as pursuit of safer and better means of surveillance whereas later developments slowly evolved to control the environment for self-advancement.

The doctrinal precepts of high ground initially led to the development of crude propellants for launching primitive military ordnances like fire pots, fire arrows, etc. The requirement to launch farther and higher led to further refinement and by the 13th century, the Chinese had developed solid propellants like gunpowder which revolutionised the delivery of military ordnance like fire-arrows, fire spears and their refined successor— the rocket—into battle. Rocket technology continued to undergo refinement over the ages until it reached a stage wherein it could launch military ordnance (missiles), platforms (satellites), and even humans on to the ultimate high ground of space.

While the Chinese pioneered the early development of rocketry, it was the Indians who first inculcated this revolutionary technology of rocketry into their military doctrines and regularly fired them in battles.[1] From the 16th to 19th century Indians used rockets more extensively than anybody else. For example, in the battle of Panipat in 1761 between the Afghans and the Marathas, the Marathas are reported to have launched barrage fires of up to 2000 rockets at a time. Rockets were also used in the battles of Srirangapatnam by the Mysorean armies in 1792 and 1799. In 1792, Tipu Sultan's rocketeer contingents used rockets against the British to devastating effect.[2] Following the rout of the British by Tipu's rocket contingents, the British studied these revolutionary weapons,[3] refined them and finally led by Colonel William Congrieve introduced their own version of rockets. These Congrieve-design rockets were later used by the British against Napoleon's French armies.[4] By the end of the 19th century, rocket technology and its related military techniques had migrated from the East to the West (where it had developed) and reached the present levels of refinement, proficiency and efficiency.

Going Higher and Farther

The West explored uses of the rocket beyond mere delivery of munitions and by 1906, a German named Alfred Maul successfully took aerial photographs of the ground by attaining cameras to solid-fuelled rockets.[5] However, these crude attempts were discontinued following the advent of airplanes by 1903. The arrival of aircrafts and refinement of artillery weaponry reduced the military utility of rockets for the time being. Aircrafts from their elevated locations enabled acquisition of high ground and its related advantages. Most early missions of aircraft were related to observation and delivery of munitions. Later technological refinement allowed missions of communications, navigation, weather observation, etc. The enabling of military advantages led to contests and battles for control of the environment of air. Apart from crude military contests for control of the environment, by World War I the French had introduced small solid-fuelled La Prieur rockets, which were designed to be fired from French or British bi-planes against German observation balloons. Rockets using a solid propellant had been used as weapons by all sides in World War I, and as a result, the Treaty of Versailles forbade solid-fuelled rocket research in Germany.[6]

Nevertheless, technological refinements continued over the years, the principal impediment to higher and farther ranges being the type of propellant in use. Efforts and experiments to develop more powerful liquid-propelled rockets capable of higher altitudes and ranges continued and by the 1930s solid-fuelled rockets were being replaced by liquid-fuelled rockets. In 1934, the Germans led by Wernher Von Braun developed the 'Aggregate-1' (A-1) rocket which was powered by a combination of liquid oxygen and alcohol. The A-1 was followed by the A-2 and then the A-3 at *Heeresversuchsstelle Peenemunde* (Army Experimental Station Peenemunde). By 1938, Germany had begun invading huge portions of Eastern Europe. Adolph Hitler recognised the need for an effective ballistic missile weapon and the German Ordnance Department assigned to the Peenemunde team the task of developing a ballistic weapon with a range of 150 to 200 miles, which could carry a one-ton explosive warhead and also be appropriately sized so as to be compatible with existing German rail and road infrastructure. These criteria led directly to the development of the A-4 rocket. Between 1937 and 1941, up to seventy A-3 and A-5 rockets were launched, each launch testing components for use in the proposed A-4 rocket. Finally, on 03 October 1942, an A-4 launched from Peenemunde following its programmed trajectory landed on its target 120 miles away, heralding the arrival of the space age. This successful ballistic launch is generally considered the precursor of modern-day ballistic missiles and space launch vehicles.

On to the Fringes of Space

By July 1943, Hitler had authorised full-scale development of A-4s, now renamed *Vergeltungswaffeswi-2* (Venegance weapon-2) and on 7 September 1944, the first V-2 rocket was launched against London. Around 4,300 V2s were launched against England, France, Belgium and other allied targets. The V-2s passed through the lower fringes of space uncontested at a speed of around 3,600 miles an hour and fell on their targets.[7] The allies had no way of intercepting them, hence the employment of space as an elevated pathway for delivery of ordnance on to the intended target became the dominant perception regarding the military utility of space. However, these perceptions were tempered by the technological difficulties of those times since assured access beyond atmosphere on to earth orbit was still more than a decade away. Nevertheless, institutional efforts to explore the feasibility of exploiting other capabilities afforded by the high ground of space were in place and by 1946, US institutions like the RAND corporation had undertaken studies which indicated that a "satellite offers an observation aircraft which cannot be brought down by an enemy who has not mastered similar techniques". They also proposed military satellite systems though the requisite technology to launch vehicles into space orbits was then not available.[8]

Into the Realm of Space

Towards the end of World War II, interest in German rocket technology was immense and thereafter, both the Soviets and the Americans began the quest for plundering German rocket expertise to develop their own programmes. Under the Yalta agreement the Soviets had been granted jurisdiction of Peenemunde, Mittelwerk plant area, Bleicherode, etc. where Von Braun's team and his V-2 facilities were located. However, the US out-raced the Soviets[9] and vide "Operation Paper clip" collected all the V-2s, V2 components, documents and technical personnel (including Von Braun and his team) that they could. By the time the Soviets arrived and assumed jurisdiction, mostly unskilled lower echelon workers were available whom the Soviets initially seized and later repatriated in 1953 since they were not of much use.[10] The Soviets nevertheless decided to undertake their quest on their own.

On 16 April 1946, Von Braun and his team launched the first V2 for the US. Over the next six years, sixty-four V2s were launched. Later it developed into the US Redstone rocket which was intended to serve both as a space launcher and as a tactical ballistic missile. The V2 was the basis of both American and Soviet missiles. The first rudimentary Soviet ICBM—the SS-1 Scud—was also nothing more than a refined copy of captured German V2s. Most such programmes were considered extensions of long-range artillery and aimed at mating nuclear weapons with long-range weapons for delivery of ordnance on to targets. In the US, missile research and development competed directly for precious funding with long-range bombers which were the conventional platforms for delivery of munitions. Initial post-war interest in long-range guided missiles soon succumbed to an Air Force policy that relied on strategic bombers carrying air-breathing missiles.[11] Owing to the intense scepticism within the US on the inability of existing rockets to deliver atomic bombs weighing 10,000 pounds, ICBM research was stopped in 1947. However, it was restarted later in 1954 when smaller and more powerful hydrogen bombs came into being.[12] By 1955, in response to the perceived missile gap between the US and the Soviets, Eisenhower directed that the Air Force's Atlas ICBM project be the nation's first priority.[13]

On the other hand, the prevailing circumstances of those times also dictated the need for both parties to stay informed of each others' military activities. This led to the belief that while entry into an adversary's airspace was violation of international law, it certainly was an endeavour worth the risk and effort to stay informed. This initially led to proposals like Eisenhower's "open skies" which were rejected by the Soviets. This in turn fuelled the quest to go higher and observe the adversary from an altitude high enough not to be shot down, leading to the development of high-altitude aerial platforms like the U-2 and later SR-71s. It also provided the impetus for space-based observation, surveillance and reconnaissance platforms. The focus soon shifted beyond delivery of ordnance on to other applications like observation, communication, etc. Based on the RAND recommendations, the US Air Force initiated Operation Feedback in April 1951. This programme researched the possibility of using satellites for military observation and other purposes. By 1954, it was converted into a weapon system plan WS-117L, which included three reconnaissance programmes; reconnaissance via recoverable film systems (CORONA), infrared surveillance for missile launch detection (MIDAS)[14] and reconnaissance via electro-optical systems. By the following year, the US Air Force had contracted the development of reconnaissance satellites.

Unlike the US, the Soviet Union did not have a huge fleet of long-range bombers, hence the prospect of ICBM development appealed to them. It did not face opposition from competing weapon platforms either. What it did have were relatively primitive atomic weapons that were bulky and required tremendous lift to propel them across an intercontinental range. Thus they proceeded to create heavy-lift launch vehicles which would allow them to deliver ordnance over long distances and thereby contain the arms imbalance caused by US's preponderance of manned bombers. By August 1953, the Soviets tested their hydrogen bomb. Drawing heavily on German expertise they made considerable progress in their missile development programmes and on 3 August 1957, the Soviets successfully flight-tested the world's first ICBM, the R-7 codenamed SS-6 Sapwood.[15] Based on their SS-6 ICBM booster, on 4 October 1957, the Soviets launched the world's first artificial satellite, *Sputnik-1* (Traveller-1), thus heralding the dawn of the space age.[16] With this launch, the superiority of Soviet military space technology was conclusively demonstrated and it was evident to the entire world that the Soviets now possessed the powerful military troika of nuclear weapons, ICBMs and satellite launchers. Soviet morale sky-rocketed while US national morale nose-dived. The Soviets added insult to injury by offering assistance to the US through the UN programme for technological assistance to primitive nations.[17] The Soviets had conclusively demonstrated that they could now use the ultimate high ground for delivery of munitions as well as for placing of military platforms like satellites for observation, reconnaissance, etc. on the highest ground possible beyond the reach of anybody else (particularly the Americans).

This caused an enormous uproar in the US and the Americans attempted to match Soviet space capabilities with the launch of their Vanguard. However, on 6 December 1957, with the whole world watching, America's Vanguard exploded on its launch pad itself.[18] The failure accelerated US space efforts and caused even steeper competition with the Soviets. Subsequent to the shock of *Sputnik,* the Americans entrusted the responsibility for military operations in space to its Air Force.[19]

The overwhelming and emergent impetus for military utilisation of space systems beyond mere delivery of munitions arose in the mid-1950s, following the failure of American high-altitude strategic reconnaissance aircraft like the U-2 flying out of Pakistan, Turkey, etc. to photograph and locate Soviet ICBM sites.[20] The shooting down of a U-2 on 01 May 1960 by a Soviet Air Force surface-to-air missile and recovery of the pilot, Francis Gary Powers and the aircraft by the Soviets precipitated matters and made the Americans suspend their U-2 programme. Increasing Soviet proficiency at aerial interception by fighter aircraft and anti-aircraft missiles made aerial reconnaissance an extremely risky proposition. Reconnaissance from beyond the atmosphere way above the reach of Soviet arsenal became a necessity, fuelling the urgency for consumption (rather than just plans and proposals) of space-based observation systems. The US's Air Force systems command had initiated a strategic satellite reconnaissance programme, code-named "Pied-Piper" way back on 16 March 1955; however, the programme struggled along under funded and was ignored until the above-mentioned incident. Thereafter the situation underwent a tremendous change and the US soon obtained photos of Soviet missile installations, this time from space. Nevertheless, space did not render aerial reconnaissance obsolete, and the U-2s and satellites were augmented by the more capable high speed, high altitude SR-71 by 1962.

On 28 February 1959, the US launched the first experimental reconnaissance satellite Corona under the name *Discoverer-1,*[21] which was followed by the world's first military navigation satellite, the *Transit-1B* on 30 April 1960,[22] the first weather satellite *TIROS-1*[23] in April 1960 and the first military communication satellite the *Courier-1B,* on 4 October 1960. The Soviets followed suit and by 1961, their first military photoreconnaissance satellite—*Zenith*—under the garb of Kosmos, was launched. Thus, the foundation for military uses of space was laid virtually immediately after it became possible to insert space craft into near-earth orbit.

From the foregoing it is apparent that as a field for military operations, space literally never existed as a real possibility prior to the twentieth century. It was in World War II that the possibility of using space as an arena for military operations was first introduced. In a fundamental sense this use of space was confined to rockets like V-2 using the lower fringes of space as a pathway to targets. However, the use of rockets was conceived initially as extension of long-range artillery for efficient delivery of military ordnance into battle. For example, the initial utility of ballistic missiles was limited to enabling delivery of nuclear weaponry across great distances in a matter of minutes unencumbered by airborne opposition. In case of the V-2, though it traversed through the fringes of space, it was not explicitly a space vehicle though its flight to target was uncontested. However, once the realm beyond atmosphere opened up, the first space-based platforms aimed at enhancing terrestrial military capabilities were soon in place.

Space Race during the Cold War

No precise date marks the beginning and end of the Cold War. By some accounts it began in 1939 following the annexation of the Baltic States by the Soviet Union while others date it from 02 September 1945 till 26 December 1991 when the Soviet Union disintegrated. Regardless of the precise time, military competition for harnessing the spoils of space and rocket technology was well established by the period following World War II and intensified as time progressed. By the mid-1950s both the Soviets and the US were engrossed in creating their respective space delivery and at lesser pace application platforms. The first few years following the opening up of the realm of space by *Sputnik* were witness to a flurry of intense military space activity. The Americans had now taken the lead and in terms of application satellites the Soviets clearly trailed behind in the first four years itself, as evidenced from Figure 1.1. What is equally significant is that within the first four years itself, almost the entire range of capabilities afforded by the realm of space for conventional military force enhancement was in place. Apart from greater space system proficiency, this was clearly indicative of the solid technological and doctrinal base of the US on the subject.

Except in the case of ELINT, where the US debuted by 1962 and the Soviets half a decade later,[24] space-based capabilities for enhancing conventional and strategic military capabilities of the US were very much in place, much to the consternation of the Soviets. Towards the end of the decade since *Sputnik* (1967), the US had completed deployment of virtually the entire range of space application programmes and was into its second- and third-generation programmes whereas by the same time the Soviets had also attained proficiency in application satellites considering that the first generation of Soviet ELINT satellites, its navigation satellites and ocean reconnaissance satellites debuted in this year.[25] The Outer Space Treaty was also inked at the same time though both the super-powers were actively pursuing their respective military agendas of controlling the realm of space with ASATs.

Number of Satellites–Year wise

Satellite	1958		1959		1960		1961		Total		First Soviet launch Date
	US	USSR	US	USSR	US	USSR	US	USSR	US	USSR	
Comm	01	00	00	00	02	00	00	00	03	00	1964
Nav	00	00	01	00	01	00	03	00	05	00	1967
Photo-Reccee	00	00	06	00	06	00	13	01	25	01	1962
Early Warning	00	00	00	00	02	00	03	00	05	00	1971
Met	00	00	00	00	02	00	01	00	03	00	1963

Figure 1.1

Early Attempts at Space Control

As in the case of aerial platforms like the U-2, the Soviets threatened to shoot down US satellites and began development of Anti-SATellite (ASAT) systems, thereby heralding the arrival of competition for control of the environment of space. In response the Americans began contemplating means of securing and preserving their military assets in space. Thus, within the first few years of the realm of space opening up all the possible space-based military missions of ordnance delivery (force-application), conventional military force enhancement/multiplication (force-enhancement) and battles for control of the environment of space (space control) were already in place.

As a matter of fact, projects aimed at denying the realm of space had been contemplated ever since it became possible to insert objects into space. For example, as early as 1958, project Argus was undertaken by the Americans to create an artificial radiation belt around near-earth by detonating a nuclear device in space. Such a detonation would disperse a cloud of radioactive particles around Earth which would interfere with radars, communications and the electronic systems of ballistic missiles and space craft. Three tests were undertaken vide this project. *Argus-1* was launched on 27 August 1958, with the weapon detonating at an altitude of 161 kilometres (100 miles). *Argus-2* followed on 30 August with the weapon detonating at 293 kilometres (182 miles). The last in the series, *Argus-3* took place on 6 September, with the weapon detonating at 750 kilometres (466 miles). However, the radiation belt did not prove strong enough to disrupt enemy ICBMs and act as a missile shield. During 1962, the Soviets were also known to have exploded three nuclear weapons in space at an altitude of around 200 miles.[26] However, the passage of the Limited Nuclear Test Ban Treaty in 1963 made such detonations in space unlawful and simple verification measures made them easily detectable.[27]

Apart from these, both the Soviets and the Americans had developed ballistic missiles to enable them to deliver nuclear and conventional weaponry across intercontinental distances. The Soviets had an edge over the Americans with respect to rocketry and missile technology whereas the Americans had an edge over the Soviets with regard to application satellites in the early years of the space age. Hence it was natural that military planners on both sides began programmes and strategies to contain their weaknesses and counter their rivals' strengths. This moulded the perceptions on the utility of ASATs and their early development in either case.

Early American efforts at countering the above capability included programmes like the Nike-Zeus, Project Defender, and the Sentinel Program. The late 1950s Nike-Zeus programme involved firing Nike nuclear missiles against oncoming ICBMs, thus exploding nuclear warheads over the North Pole. This idea was soon scrapped and work began on Project Defender in the 1960s. Project Defender attempted to destroy Soviet ICBMs at launch with satellite weapon systems which orbited over Russia. This programme proved unfeasible with the technology from that era. Work then began on the Sentinel Program which used anti-ballistic missiles to shoot down incoming ICBMs. The technology in case of Sentinel was also yet to mature, nevertheless deployment was limited by the ABM treaty of 1972.

In contrast the Soviet effort was primarily aimed at containing the threat from US observation and reconnaissance satellites. As in the case of the U-2s, the Soviets were determined to prevent orbital spy missions and by 1959 they had conceived the idea of *Istrebitel Sputnikov* (Satellite Destroyer). In spite of the rush by both superpowers to develop ASAT weaponry, until 1963 neither had succeeded in acquiring an operational system capable of actually intercepting satellites.[28] By the following year the Soviets expanded their PVO-Strany air defence branch to air and space defence PKO *Protivo Kosmichesokaya Oborona* aimed at repelling attacks from space. However, actual flight tests were conducted only in 1967–1968.[29]

From the foregoing, it is apparent that space was being exploited for almost every conceivable military mission since the dawn of the space age, though the primary contribution of space in conflicts was limited to threats, counter-threats and deterrence rather than raining down actual punishment from the heavens. Nevertheless, what actually limited this mission were the technological and economic challenges of those times rather than legalities or ethical considerations.

Early Role in Conflicts

Strategic Uses

The early role of satellites was perceived to be mainly for nuclear dissuasion. Their primary role was to provide information about the launch of ballistic missiles. Space-based systems overcame the obstacle posed by the Earth's curvature and were able to double the warning time as compared to Earth-based radar systems. This led to perceptions on the strategic utility of space-based systems.

In the half decade following the launch of *Sputnik,* while most of the military uses of space were explored to some extent, actual military use was largely limited to sabre-rattling by both sides. It was during the Cuban missile crisis that the strategic utilities of space-based observation became demonstratively apparent. During 1963 space systems played a tremendous supporting role in the Cuban missile crisis. Although they did not locate missiles in Cuba, US *Discoverer* satellites gave their military and political decision makers reasonable estimates of Soviet nuclear delivery capabilities in terms of aircrafts and ICBMs. This conveyed to them that the capabilities of Soviet nuclear forces were quite limited. Knowledge of this threat enabled Kennedy to call Khrushchev's bluff. On the contrary, Soviet counterpart systems like Zenit told Khrushchev that the US was actually positioning forces to attack Cuba and that the US Navy was moving into position to stop Soviet ships. The Soviets backed out, and the crisis was averted.

This incident soon led to belief in some quarters on the use of observation satellites for promoting international security by reducing the risk of accidental war and pre-emptive strikes. It brought into focus the role of space 'systems for monitoring military activity, providing early warning to reduce the likelihood of surprise attack, and serving as National Technical Means of Verification (NTMV) to enable and enforce strategic arms control. It led to the development of the "Sanctuary" doctrine in the early years which was aimed at using space surveillance systems to make nuclear wars less likely.[30]

Strategic to Tactical use

Following the build-up of nuclear and conventional arsenal amongst the space powers and the consequent concerns of the world, which in turn led to the formulation of legalities and conventions, the actual scope of a conventional total war amongst the two super powers was bleak. Mutual antagonism never got reduced to military confrontations of the World War II type. While both avoided direct confrontation and mass destruction, they nevertheless concentrated on expanding their spheres of influence around the globe, in the process promoting local conflicts wherein their role was limited to that of indirect participants but one which they pursued with extraordinary vigour. In brief the spectrum of conflict had evolved beyond massive deployment and nuclear showdowns to localised low-level conflicts under the indirect patronage of the super powers in most cases. In such a scenario the tactical utility of space-based assets in terms of observation satellites for monitoring military activities like movement of troops, armoured columns, the role of ELINT satellites in assessing the enemy ELINT ORder of BATtle (ORBAT) as well as of Ocean Reconnaissance satellites in detecting and tracking multiple targets in the ocean both sub as well as surface soon became more prominent. This led to the initial shift from the pre-eminence of strategic priorities for use of space assets, toward a balanced approach that considered tactical requirements equally valid. It soon became apparent that satellites like aircraft were complex instruments of warfare which had applications at strategic, operational and tactical levels; hence confining the use of space to strategic applications alone would amount to gross underutilisation of scarce and costly resources. Thus, practical requirements of the tactical battlefield soon started taking precedence over strategic reconnaissance aimed at containing the threat of ballistic missiles or accidental nuclear launches. This is evidenced by the fact that the pre-eminent role played by satellites for photo-reconnaissance of ballistic missile and nuclear sites soon gave way to satellites aimed at promoting tactical objectives like reconnaissance missions for observation of tactical battlefield, communication satellites for battle field communicability, and navigation satellites for targeting.

It was the tactical requirement of guidance of anti-ship cruise missiles which prompted the development of ocean-reconnaissance satellites. By the end of 1950, the Soviets had developed cruise missiles which could strike beyond the range of radars installed on the launching ship. The requirement of a new guidance system with ranges much more than conventional radars led to the development and deployment of *Upravlayemi Sputnik* (Controlled Satellite) by 1967 which used radars to locate vessels in the ocean.[31] Based on the lessons of the Cuban missile crisis, the Soviet naval forces expanded immensely and by the early 1970s the Soviet anti-ship cruise missile, supported by a seemingly well-integrated ocean surveillance system, came to pose a threat which US units were not prepared to counter.[32] The US responded with their ocean reconnaissance satellites almost a decade later. However, present-day media reports reveal that the US plugged this lacuna in those times by launching "POPPY" satellites in an ELINT role to collect radar emissions from Soviet naval vessels and support ocean surveillance. The programme operated from December 1962 through August 1977.[33] The 1970s also witnessed the growth of SIGnals INTelligence (SIGINT) systems like Rhyolite, Chalet, Magnum, etc. which were launched by the US to monitor missile telemetry as well as for communication surveillance between Soviet military commanders in terms of telephone calls, walkie-talkie traffic, etc. Certain reports indicate that such satellites were also used for general espionage in Asia during the 1971 Indo–Pak war in Vietnam and later to spy on China.[34] Thus the period witnessed the use of SIGINT satellites for tactical applications like monitoring thousands of telephone calls at a time, monitoring of walkie-talkie traffic as well as strategic surveillance related to missile development and deployment.

Of particular interest is the increasing role of observation satellites which was demonstrated time and again during numerous conflicts ranging from the Sino–Soviet border skirmishes over Damansky Island and later Goldinsky Island to full-scale wars like the Indo–Pak war of 1971,[35] the 1973 Middle East conflict, the 1982 Falkland wars and every major conflict till the recent space war. For example, during the Indo–Pak war of 1971, the Soviets launched high-resolution photo-reconnaissance satellites like *Kosmos 463* and *464* so as to enable observation of the areas of interest each day near noon on the eastern battlefield. This played a seminal role in effective decision making thereby influencing the character of the battle.

Similarly, during the Middle East conflicts of 1973, the Soviets launched six photo-reconnaissance satellites during the 21 days of hostilities. Three days prior to the outbreak of hostilities on 06 October, the Soviets launched their first satellite for surveillance of the area. Within an hour of the Egyptian attack against the Israelis across the Suez Canal, the Soviets had launched a high-resolution photo-reconnaissance satellite which not only recorded activity on the Egyptian front, but photographed the Syrian attack against Israeli positions on the Golan Heights. During the next three weeks the Soviets constantly monitored the situation.

It was through satellite photographs that the Soviets convinced President Sadat that the Israelis had made serious incursions across the canal and additional Soviet military aid was required. Later, when the Egyptian 3rd Army was cut off from Cairo and the Egyptian military situation continued to deteriorate, satellite photographic coverage provided confirmation of the situation. Satellite photographs were used by the Soviets to press Sadat to agree to a ceasefire. There is little doubt that the Soviets used their photo-reconnaissance satellites effectively to monitor the Middle East war and to persuade Sadat to take certain actions. However, the Soviets appeared to use their capability only as a diplomatic tool and not as an aid to the Egyptian military commanders.[36] Thus by the early 1970s itself, the decisive role of observation satellites in conflicts and wars was established. The seminal role of reconnaissance satellites in military conflicts can be gauged from the fact that historically they have always constituted the largest category of military satellites, from a purely numerical point of view. As a matter of fact, while the US had an early lead from the 1960s through the early 1970s with a launch of more than 140 satellites during the period,[37] Soviet satellite surge capability soon overtook them. The Soviet Union and Russia together have launched over 780 photo-reconnaissance satellites.[38]

Enhancing Military Force Capabilities

By the late 1970s experimental space missions evolved into full-fledged military missions with dedicated constellations aimed at enhancing the military capabilities of conventional forces. With every passing conflict, the missions continue to undergo greater refinement and a number of experiments also grew into major satellite programmes. Apart from the aforementioned examples on the use of space by the Soviets, the Americans also used space systems in operations like Urgent Fury (Grenada), El Dorado Canyon (Libya), Just Cause (Panama), etc. However, the employment was incomplete and often ad hoc. That is, only a subset of the full range of space systems was used. Moreover, the individual commander's knowledge of space often determined the employment of space capabilities. Despite some shortcomings, the operations in Grenada, Libya, Grenada, and Panama were key milestones in space operations and contributed to knowledge of the employment of space capabilities.[39] Nevertheless, during the Cold War era, space assets were generally used only episodically in local wars and armed conflicts. They were also largely limited in their mission to particular roles like the presence of certain reconnaissance satellites in orbit ELINT, SIGINT satellites providing intelligence inputs, or weather information from a particular satellite being available or not.

Military Use of Space in Recent Conflicts

Operation Desert Storm

Following the invasion of Kuwait by Iraq in August 1990, the Americans initially launched Operation Desert Shield to protect states like Saudi Arabia from further Iraqi aggression and followed it up with Operation Desert Storm in January 1991, whereby the US and its allies decimated almost the entire Iraqi military apparatus. What is of interest is that even before the shield (of air, naval, land forces) was in place, space systems were already in orbit over the area of interest.[40] Although these assets played only a supporting role in the Allied build-up and actual combat which later followed, what distinguished them from their role in previous conflicts was that for the first time the entire range of space assets was employed in direct, though not fully integrated support of combat operations at all levels. The contribution of space assets to the unqualified military success of the coalition was considered so critical that the conflict was labelled the "first space war"[41] and the appellation stuck ever since. The experiences and lessons of this war not only brought into the limelight the seminal role of space in decisively influencing the outcome of conflict but also shaped global perceptions on the subject as it was the first real demonstrative applied experiment wherein space-based assets decisively enabled the attainment of total military superiority. Elaborate, exhaustive treatises on the vital contribution of space assets to the military effort are available in open literature, hence the intent is not to recount the same but to briefly encapsulate the salient aspects and comprehend the enormous force-multiplication enabled by space-based capabilities.

Desert Storm was the first war when the entire range of satellites for fulfilling roles of force enhancement like Intelligence Surveillance Reconnaissance (ISR), navigation, meteorology, communications, Early Warning (EW), etc. were available to the coalition forces. It was the first occasion when the cumulative effect of over sixty satellites[42] dedicated to the coalition effort was demonstrably evident in influencing the outcome of war. The information thus provided ranged from Multi Spectral Imaging (MSI) satellites providing a celestial view of the battlefield to navigational satellites providing precise targeting, safe manoeuvring, precise munitions delivery information and so on. Dedicated Met-Satellites provided weather inputs and supported the entire spectrum of combat operations. They provided near real-time information critical to campaign planning on the ground, sea and air. Early Warning (EW) satellites like the Defence Support Program (DSP) basically designed to provide EW of Soviet intercontinental missiles were instrumental in detecting and shooting down incoming Scud missiles. Communication satellites provided reliable and near-total intra- and inter-theatre communication links enabling battlefield communications as well as communications between the Iraq-based coalition commanders and the US White House.[43]

As the war progressed, the burgeoning demand for communication satellites made the coalition forces lease commercial space systems to cope with increasing military demands. Coalition communication systems were reported to carry more than 700,000 telephone calls and 152,000 messages per day during the most intense part of Desert Storm.[44] Out of this, satellite communications systems carried 85% of the total inter-and intra-theatre load.[45] Similarly, the demand for GPS systems also outstripped military demand.[46] Apart from the fact that the entire constellation was not available, leading to reduced capability,[47] it was only after the first troops arrived in Saudi Arabia that the vital need for GPS receivers in the desert conflict was fully appreciated.[48]

While there can be no contesting of the decisive contributions of space in enabling the coalition victory, what is glossed over in all the hype surrounding the "first space war" is the fact that in this war technology was clearly ahead of prevalent doctrines. With regard to the coalition's use of space assets, there existed a serious doctrine–technology mismatch. The doctrinal utility of space systems was evidently not very well known to the coalition forces, resulting in sub-optimal utilisation and integration into war-fighting.[49] Apart from utility awareness which was demonstrably lacking, the demands on space systems were also not anticipated (or the potential to contribute underestimated), resulting in knee-jerk acquisition responses. Acquisition of space systems was based on increasing exploitability awareness as the war progressed and not as a deliberate planned process. The saving grace was the fact that the Iraqis had extremely rudimentary knowledge and even more rudimentary capability (close to nil), allowing the coalition a space "walk-over".[50]

Kosovo and Kargil

The end of the millennium was witness to two major conflicts on the globe—one in the Serbian province of Kosovo and another on the icy heights of Kargil. The Kosovo conflict witnessed the largest armada of spacecrafts ever brought to bear upon a single war in the history of the millennium. In contrast the Kargil conflict was witness to not a single spacecraft being brought into the conflict although India was an established space power by then. The contribution of space-based assets both in the conduct of military operations as well as influencing the outcome of war is instructive. While Kargil was characterised by lack of information in all aspects ranging from intelligence on enemy locations to targeting information, weather inputs, etc. Kosovo was characterised by a surfeit of space-based military information for the coalition forces which paved the way for nuanced application of military power and consequently decisive success in battle.

Certain pertinent issues with regard to the utility of space-based assets need to be examined in this context. Had India harnessed its extant space capabilities for national security, could it

- Have detected the enemy incursions?
- Assessed the scale of incursions more accurately?
- Delivered its fire power and applied air power more precisely and persistently?
- Finally, have lost less in terms of human lives and limbs?

The answers would be apparent even to the uninitiated. The induction and integration of space-based capabilities into conventional military capabilities is inescapable, especially so if extant capabilities are available.

Operation Iraqi Freedom

Desert Storm was referred to as the "first space war" because every aspect of military operations depended, to some extent, on support from space-based systems. Space systems then were used 'for force enhancement by harnessing military space capabilities of position/navigation, weather, communications, imagery and tactical early warning of missile attack, etc. Space systems in Operation Iraqi Freedom also performed similar functions; the primary difference being that refinements in user technologies and doctrines on optimal exploitation of space assets allowed greater integration of space with the conventional war-fighting apparatus. Unlike in Desert Storm, space-based capabilities were integrated with actual military operations at all levels and the incremental gains in combat effectiveness were demonstratively apparent to the world. The success of Operation Iraqi Freedom depended heavily on improved support and force enhancement capabilities provided by space-based assets. The improvements over what was accomplished during Desert Storm are compelling. The whole intent was focussed on bringing about an integrated effect to the battle space by putting air and space together and supporting troops in the field as well as other service components.[51] The cumulative effect of the above integration is apparent when considering the fact that it was this combination of space as well as aircraft, precision weapons and updated intelligence which allowed the performance of such previously improbable feats as destroying only the fourth through seventh floors of a building.

Notwithstanding the above, it needs to be borne in mind that ever since the demise of the Soviet Union and the end of the Cold War, most conflicts involving the use of space have been unequal contests which pitted the enormous capabilities of the US against contestants with nascent or at times even nil space capabilities. The lessons consequently need to be applied with caution; even without space capabilities, enormous asymmetric advantage has always been with the US. However, in the absence of any other battlefield experiences, the same are instructive. The most persistent lesson manifest in the successive conflicts wherein space-based assets have been used is the asymmetric advantage of information dominance enabled by space assets. The other outstanding aspect is that unlike most military systems which demand a numerical increase with every passing conflict to maintain advantage, space systems are characterised by relatively stagnant or modest increases in numerical terms.[52] Not only does it not demand an increased numerical count, it reduces the numerical demands of other military systems like aircrafts, munitions, and a host of other military systems. This is validated by the fact that with respect to Operation Desert Storm, Operation Iraqi Freedom was witness to a modest increase in satellites because the technological and doctrinal aspects on use of space assets witnessed an increasing refinement. With regard to the cost-economy factor, this indicates that while initial costs and investments may be high in building space capabilities, the recurring costs would be much less. For example, during Operation Iraqi Freedom, while the increase in the number of satellites was marginal, US military forces capitalised on available space-based assets, melding revised doctrines and advanced technology into strategies which paid high dividends on the battlefield.[53] While the coalition air forces used up to 2,500 aircraft in Operation Desert Storm, in Operation Iraqi Freedom only 1,900 were employed.[54] While 542,000 coalition military personnel participated in Operation Desert Storm, only 350,000 personnel were used during Operation Iraqi Freedom.[55] Even with regard to munitions, as against 10% precision-guided munitions in Operation Desert Storm, in Operation Iraqi Freedom up to 68% were precision guided,[56] leading to lesser wastage of munitions and lesser collateral damage. The point is that, apart from their other virtues, use and integration of space into conventional military capabilities make enormous economic sense.

Conclusion

The role of space has witnessed an expansion with every passing conflict and would only expand further as technology and doctrine nature, enabling acquisition of greater military advantage. Increasing proficiency in cheaper and smaller micro and nano-satellites would enable greater expansion of their role in influencing terrestrial war-fighting in addition to providing an operational responsive space capability in times of wars and crises. As mentioned previously, increasing utility of space in influencing military results has also led to an increasing need for the capability to preserve and protect assets in space. This has given rise to fears in the US that an adversary could turn the asymmetric advantage provided by space systems into an asymmetric vulnerability if space superiority cannot be maintained.[57] As a consequence, the US as in the early years of the space age has restarted the quest for controlling the environment of space. The essential difference this time is that it is the lone player and does not have to contend with a rival super power like the Soviet Union.

Notes and References

1. For a more complete account of the early development and use of rocketry, see Cliff Lethbridge "History or Rocketry" Chapters 1–6, at http://www.spaceline.org/history/1.html

2. Two of the rockets fired by Indian troops in 1792 are on display at the Royal Artillery Museum in London. One of these rockets is made up of an iron case 10 inches long by 2.3 inches wide. It is bound to a metal sword that is 40 inches long. The second has an iron case 7.8 inches long by 1.5 inches wide bound by leather strips to a bamboo stick that is 6 feet, 3 inches long. Each rocket is thought to have a maximum range of 1,000 yards, and eyewitness accounts in 1792 indicated that just one rocket killed three men and injured four others. For details, see ibid, Chapter 2.

3. Certain reports indicate that in 1770, a British Captain named Thomas Desaguliers examined rockets brought from India in the Royal Laboratory, Woolwich, England, but failed to reproduce reported range or accuracy. (Some would not even lift from their stands). See www.russianspaceweb.com/rockets_pre20th_cent.html

4. Calvin J. Hamilton, "A Brief History of Rocketry," courtesy KSC/NASA at http://www.solarviews.com/eng/rocket.htm

5. Ibid.

6. See "Militarisation of Space," *Wikipedia* at http://en.wikipedia.org/wiki/Militarisation_of_space.

7. However, for the Germans, the V-2 offensive came too late to influence the outcome of the war. By April 1945, the German army was in full retreat everywhere and Hitler had committed suicide in his bunker in Berlin.

8. "Preliminary Design of an Experimental World-Circling Spaceship," Santa Monica: RAND Corporation, 1946. The above was followed by many suchlike reports which went into considerable details regarding the engineering and utility of space-based observation for intelligence, weather, etc.

9. In the case of Bleicherode, the US Army left the town on 30 June 1945, a day prior to the Soviets taking over the area. Ref "German Legacy in Soviet Rocketry" at www.russianspaceweb.com/rockets_ussr_germany.html

10. The importance of Von Braun to both the Soviets and the Americans can be gauged from the fact that the Soviets in a later futile attempt to recruit Von Braun sent an emissary across into the US zone. The emissary however was quickly intercepted by the US military and escorted back empty-handed.

11. Major Ramey, "Armed Conflict on the Final Frontier: The Law of War in Space," Ch. 2, *Air Force Law Review,* 13 March 2001, at http://www.space4peace.org/slaw/lawofwar.htm

12. Major Michael J. Muolo, "Space Handbook," Vol. 1, Ch. 1, pp. 3–4, *Air University Press,* December 1993.

13. The US Army had been assigned the task of developing the Redstone rocket as a tactical ballistic missile; the Air Force was assigned Atlas ICBMs and the Navy sea-launched rockets like the Vanguard.

14. This system was later replaced in the early 1970s by gesynchronous satellites of the Defense Support Program (DSP) which were considered to be highly effective in offering notice of a missile attack within moments of launch.

15. Curtis Peebles, "Battle for Space", Beaufort Books Inc., New York 1983, p. 52.

16. The Soviets led the space race for an entire decade in terms of launches though the Americans had the edge with regard to application satellites. For a comprehensive list of the number of 'Firsts' notched by the Soviets see *Wikipedia* "Soviet Space Programme" at http://en.wikipedia.org/wiki/Soviet_space_program

17. Fred Reed, "The Day the Rocket Died," *Air and Space Smithsonian* 2, No. 4 (October/November 1987, p. 52.

18. Finally, on 31 January 1958, the first US satellite *Explorer-1* went into orbit.

19. By 1961, the US Department of Defense had entrusted the entire mission of managing and operating US military space launch vehicles and satellites to the Air Force.

20. Walter A. McDougall. "The Heavens and the Earth: A Political History of the Space Age," New York: Basic Books Inc., 1985, p. 219.

21. However, actual space-based reconnaissance began only on 18 August 1960 with *Discoverer-14* which was the first satellite to carry cameras and bring back pictures. The first flight of *Discoverer* had failed and the remaining twelve flights went awry between 28 February 1959 and 29 June 1960.

22. *Transit-1A* did not make it to orbit in September 1959 and failed whereas *Transit-1B* made it safely to low Earth orbit.

23. TIROS stood for Television and Infrared Observation Satellite.

24. While the first US ELINT satellite was launched by 1962, the first generation of Soviet ELINT satellites under the garb of Cosmos 148 appeared only in 1967.

25. For a comprehensive description of Soviet military satellites, see Brian Harvey, "Russia in Space: The Failed Frontier?" Praxis Publishing Ltd, Chester, UK, 2001, Ch. 4, pp. 120–130.

26. Curtis Peebles, "Battle for Space," Beaufort Books, New York, 1983, pp. 95–97.

27. The Americans launched the "Velta Hotel" series of satellites in 1963 and 1964 to scan above the horizon and detect nuclear tests in space.

28. Nicholas L. Johnson, "Soviet Military Strategy in Space," Jane's Publishing Company Limited, London 1987. Ch. 4, p. 139. Secondly, with regard to the US, their satellite INTerceptor (SAINT) ASAT programme was cancelled by 03 December 1962.

29. Ibid., pp. 140–141.

30. Sanctuary doctrine is closely linked to deterrence theory and the assumption that no meaningful defence against nuclear attack by ballistic missiles is possible. The advocates of the doctrine believe that over flight and remote sensing enhance stability and that space must be kept a weapons-free zone to protect the critical contributions of space surveillance systems to global security. For details, see Lt Col Peter L. Hays, "Current and Future Military Uses of Space," Conference on Outer Space and Global Security, 26–27 November 2002 at www.ploughshares.ca/CONTENT/ABOLISH%20NUCS/OuterSpaceConf02/HaysConf2002.html

31. Anatoly Zak, "Space craft: Military: US-A/P" at www.russianspaceweb.com/us.html

32. US Naval Historical Center, "From the Sea to the Stars" at www.history.navy.mil/books/space/Chapter4.htm

33. A total of seven POPPY satellites were lofted into space from 1962 to 1971: 13 Dec. 1962, 15 June 1963, 11 Jan. 1964, 9 March 1965, 31 May 1967, 30 Sept. 1969, and 14 Dec. 1971. See Leonard David, "Cold War spy satellite program declassified." Space.Com, 16 Sept. 2005 at url http://msnbc.msn.com/id/9370720

34. Australian Anti Bases Campaign, "National Missile Defence: Background on Pine Gap" at www.anti-bases.org/nmd/pinegap.htm

35. See Nicholas L. Johnson, "Soviet Military Strategy in Space," J Publishing Company Limited, London 1987, Ch. 3, p. 87.

36. Colonel Garry A. Schnelzer, USAF, "Antisatellite Weapons: Are they a valid bargaining chip?" Ch. 2, p. 19, Strategic Studies Report Abstract, April 1985, The National War College.

37. Pamela Feltus, "Aerospace Power and the Cold War," US Centennial of Flight Commission at www.centennialofflight.gov/essay/Air_Power/cold_war/AP34.htm

38. Brian Harvey, "Russia in Space: The Failed Frontier?" Ch. 4, p. 109, Praxis Publishing Ltd, Chichester, UK, 2001.

39. Lt Gen Thomas S. Moorman, Jr, USAF, "Space: A New Strategic Frontier," *Airpower Journal,* Spring 1992.

40. For example, some 15 individual US, French and British military communications satellites were already in geostationary orbit whose terrestrial footprint covered the Gulf area. See Sir Peter Anson BT and Dennis Cummings, "The First Space War: The Contribution of Satellites to the Gulf War" *RUSI Journal,* Winter 1991, p. 45.

41. It was the US Air Force Chief of Staff General Merill McPeak who used the term for the first time and the characterisation has stuck and continues to be in vogue.

42. Sir Peter Anson BT and Dennis Cummings, "The First Space War: The Contribution of Satellites to the Gulf War" *RUSI Journal,* Winter 1991, p. 45.

43. Steven Lambakis, "Space Control in Desert Storm and Beyond," *Orbis,* Summer 1995, p. 419.

44. Lt Col Steven Bruger, USAF, "Not Ready for the First Space War—What about the Second" *Naval War College Review,* Winter 1995, p. 75.

45. Ibid., p. 76.

46. During the early days of coalition buildup in Saudi Arabia, only a few hundred GPS receivers were in-theatre. The demand particularly by the US Army outstripped normal production and even resulted in soldiers writing contractors directly for the small GPS lightweight receiver. By the end of the war 4,500 receivers were in use. Lt Gen Thomas S. Moorman, Jr, USAF, Space: A New Strategic Frontier, *Airpower Journal,* Spring 1992.

47. At the start of Desert Shield, only 13 satellites were available with a further spacecraft rushed into orbit in August 1990, Anson and Cummings, p. 50.

48. Ibid., p. 50.

49. The fact that urban legends regarding GPS receivers like "I didn't need any

of that space stuff. This little box told me exactly where I was" proliferated since the 1991 Gulf War are demonstrative of the extremely low equipment awareness levels amongst the coalition troops. See Figure 1, Robert Kehler, "Space-Enabled Warfare," *RUSI Journal,* August 2003, p. 68.

50. The Iraqis had no military space assets of their own. They only had access to civil international networks like Intelsat and Inmarsat and a share in two regional telecommunications satellites operated by Arabsat. However, the Arabsat transmitting and receiving station was an early victim of the coalition bombing campaign.

51. William B. Scott and Craig Covault "High Ground Over Iraq," *Aviation Week & Space Technology,* 9 June 2003, p. 45.

52. For example, during Iraqi freedom, satellite systems of 1970s vintage like the DSP were also used as were other assets like GPS satellites which had finished their design life and were over 14 years old.

53. William B. Scott, "Space Pays Dividends," *Aviation Week & Space Technology,* 9 June 2003, p. 52.

54. Michael Knights, "Iraqi Freedom displays the transformation of US airpower," *Jane's Intelligence Review,* May 2003, p. 16.

55. Rudi Williams quoting US Navy Vice Admiral Arthur K. Cebrowski, "Transformation Director Says Cold War Space Approach Must Change," *American Forces Press Service,* 31 March 2004, site of US Department of Defence www.defenselink.mil/news/Mar2004/n03312004_200403319.html

56. Perry Fenerstein, "Generalised Representation of Space-based platforms for various orbit types," p. 28, Military Ground and Aerospace Simulation Symposium at www.scs.org/scs archive/get Doc.cfm?id=1611

57. Statement of Arthur K. Cebrowski, Director of Force Transformation, Office of the Secretary of Defense before the Subcommittee on Strategic Forces, Armed Services Committee, US Senate, 25 March 2004. Available at http://armed-services.senate.gov/statemnt/2004/March/Cebrowski.pdf.

Chapter 2

Civilian Uses and Applications of Space

Rocket technology of the 20th century heralded the arrival of the space age. The prevailing geo-politics of the era encouraged the development of military rockets that enabled mankind to foray beyond the gravitational envelope of Earth. This venture beyond Earth's planetary cradle enabled greater comprehension of the opportunities as well as challenges afforded by the realm of space for military and civil development. While most initial efforts were primarily driven by military needs, the potential for civilian applications to further the cause of human development, social welfare, science and commerce soon became demonstratively apparent. The quest to go higher and higher, into the realm of space for delivery of munitions enabled discovery of launch technologies to explore the vast expanses of the universe whereas the quest for space-based military observation provided unparalleled information and opportunities for civilian growth and development. Primarily, outer space afforded opportunities in the areas of space technology and system applications, scientific development, space launch vehicles and other technological advancements for furthering the cause of human development in the broader sense, and nation–state development in a narrower context.

The quantum of information enabled by space and computers in the 1960s drastically revolutionised life, development and growth. Elements of space-based information like Earth observation enabled greater comprehension of the environment and its complexities, space-based communications began enabling global connectivity driving other revolutions in its wake, navigational satellites began providing better guidance, navigation and collision avoidance, etc. at sea, air and land, weather satellites enabled more reliable and prompt forecasting. Apart from the now famous Revolution in Military Affairs (RMA), an equally powerful though less famous revolution in civilian affairs was also impacting and changing day-to-day life. Broadly, the revolution impacted human development in the following manner.

Revolution in Civil Affairs

Military rocket technology enabled mankind to break through the twin barriers of gravity and atmosphere. Ever since the first *Sputnik,* up to 12 men have walked on the moon and there have been up to 432[1] space farers. The ability to launch into space enabled observation and collection of data both on Earth resources as well as on the physical and scientific characteristics of the universe. This enabled access to enormous information, the diverse collection, application and analysis of which has led to a revolution in day-to-day civilian affairs. Space technology supports a variety of activities ranging from profit-driven commercial ventures to benign welfare measures in the service of all mankind and scientific pursuit. The opportunities afforded by the information endowed by space technologies are enormous. Exploitation and application of such information to human well being would be limited only by the ingenuity, imagination and dedication of the service providers and the user segments. Widespread innovation amongst vendors, service providers and users has resulted in new and extended opportunities which serve to revolutionise life on Earth. The primary drivers of the revolution are briefly dwelt upon as follows.

Revolution in Communications

Following the launch of the *Sputnik,* the Americans presaged the revolution in modern communications with the launch of Project SCORE[2] on 18 December 1958. The experiment did little apart from fulfilling its primary mission of placing the body of an *Atlas-B* missile in LEO and transmitting from the missile's payload compartment a pre-recorded Christmas message of President Eisenhower. Nevertheless, it opened a window of opportunities and possibilities which revolutionised the concept of communications. Actual applications followed the concept. By 1960, NASA had successfully launched the world's first communication satellite, the *Echo-1* experimental satellite. This was a passive communication satellite that reflected radio signals sent from the US to the UK. NASA followed it up with the first active direct relay communication satellite, the *Telstar* in 1962. Within the next four years NASA had achieved the technical maturity to launch communications satellites like the *Syncom-3,* which enabled live coverage of the 1964 Olympics across the Pacific Ocean from Japan to the US. The age of instantaneous space-based communication had arrived, heralding a revolution in the manner in which communications would serve humanity.

By 1965, the Soviets also launched their first communications satellite, the *Molniya-14* in an especially high elliptical orbit known as the 'Molniya orbit' to enable communicability across the Russian landmass and allied states. By 1970, other individual nations had also started launching their own communication satellites.

During the following decades, technological refinements allowed exploitation of communication satellites for multiple purposes. Apart from conventional uses, they also enabled transmission of good quality television signals to small, inexpensive ground receivers, thereby promoting the spread of information, entertainment, education, etc. even in remote inaccessible areas. Increasing users' demands drove the development of new technologies for better communications. For example, in the 1990s the demand for personal communication systems led to the development of new technologies to use space better. In contrast with the early satellites that could provide communicability only in terms of hundreds of voice circuits, modern communication satellites carry and handle thousands of calls simultaneously. Today, communication satellites carry about one-third of voice and essentially all international television traffic. Over the last decade, the number of TV channels broadcast by satellites has increased more than tenfold, from less than 800 in 1991 to 9,300 in 2001.[3] Technological advances in video compression and data protocol enhancement technology have made previously expensive satellite communications services such as digital direct broadcast satellite (DBS), digital direct-to-home (DTH), and Internet Access available at a lower cost.

What makes space-based communications more appealing than conventional communications is their rapid connectivity, reach as well as the ease with which they can be adapted to fulfil a variety of needs. Their inherent flexibility and adaptability enables point-to-multipoint broadcasting and also enables the unique quality of handling Internet, broadcast, telephony and corporate network traffic at the same time.[4]

They are also faster to install, can be operational in a much lesser time frame than landlines and reach places where landlines are not feasible. As a matter of fact, they enable communications on land, sea or air unhindered by geographical or topographical limitations. Laying conventional fibre optic lines in mountainous and other hostile terrain presents formidable challenges, which explains the appeal of satellite communications even in mountainous terrain, the high seas and even on airborne platforms like aircrafts. Large continents with sparsely populated swathes of land and inhospitable terrain to cover would also need to rely on satellites to relay broadband signals. Many developing nations are presently driving the demand for mobile telephony leapfrogging traditional landlines. For example, nearly all of Africa's international bandwidth is provided by satellite. African countries have a very high dependency on satellite, and in the majority of these countries more than 95% of international traffic is carried by satellites.[5] As of December 2001, Africa had an estimated 30 million mobile subscribers and 20 million fixed-line subscribers.[6] With the emergence of broadband, satellite communications have penetrated every nook and corner of the globe connecting people virtually anywhere, enabling contact, business, commerce, welfare and a host of other facilities to the content of even solemnisation of weddings on the Internet.[7] The extent of user-enabled expansion can be gauged from the fact that Internet usage in Asia–Pacific increased by 300% in 1999. In September 2000, there were about 90 million users. In February 2001, the number reached nearly 105 million and by 2005 is expected to climb to 190 million, nearly 25% of the users worldwide.[8] By 2005, China is expected to surpass the US as the largest Internet market in the world. The overall potential commercial Internet market in China is considered to be between 250 and 350 million users. India is also a fast-growing market with 38.5 million users as of November 2005 and is expected to grow in excess of 100 million in the next two years. The Internet users' base actually registered a growth of 54% in 2005.[9]

Commercial Space Communications

The sheer efficacy of space-based communications drove the spillover of space-enabled communication beyond the military and state dominion into the commercial sector. At the urging of the US, in 1964 telecommunication agencies of 11 nations[10] formed an inter-governmental consortium called the INTELSAT to coordinate satellite communications and provide international broadcast services. The following year INTELSAT used NASA facilities to launch *Early Bird-1*, the first commercial communications satellite. Towards the end of the decade INTELSAT had expanded to provide telecommunication services across the globe. In 1976, MARISAT communications satellites aimed at providing services to maritime customers entered the market. The commercial feasibility of the projects was validated by the growth and expansion that followed. For example, INTELSAT expanded to include over 100 members and provide services to over 600 Earth stations in more than 149 countries, territories and dependencies.[11] Industry estimates expect the trend to continue and the space-based information and communication market is expected to grow from around $40 billion to $120 billion or more per year by the end of the decade.[12] Nevertheless, the usual commercial market dynamics also impact upon and influence the stability of such projects. Iridium, for example, had gone bankrupt, and the economic slump of the 1990s caused an economic downturn. The slump was caused by a combination of competition from an over-saturated terrestrial landline market and technological advances like data compression. A burgeoning government and military sector demand, however, arrested the downturn. The increased tempo and wide geographic scope of the US and coalition military activities reversed the downturn into a surge in commercial leased satellite usage. The fact that a typical military communications satellite costs roughly $20 million to build, launch and operate as opposed to a commercial lease of roughly $2 million, actually served to increase the appeal of leased capabilities.[13] As regards commercial systems supporting military capabilities, it would be pertinent to note that the demand for satellite communication increased from 100Mbps for 500,000 troops during Operation Desert Storm to 700 Mbps for 50,000 troops during Operation Enduring Freedom. Out of this, 60% of the satellite communication was provided by commercial systems.

Industry experts in the present expect the additional demand for video, voice and data services to contribute steadily to market growth for the next seven to eight years. In spite of the market dynamics, the questionable fact is that space-enabled communications has revolutionised life on Earth and has expanded commercial activity and gain as never before. The demand for commercial satellite capacity globally continues to grow.

Revolution in Earth Information

Satellite-enabled information has also brought about a revolution in the manner in which the resources of the earth are utilised, managed and distributed. This revolution is driven primarily by advances in space-based remote sensing, GIS and navigation, cartographic applications, weather information, etc. Space technology has not only revolutionised perception on the environmental characteristics of the earth but provided significant technological means to understand the dynamics of the earth and the universe, assess the resources as well as the damage inflicted on the environment as also the know-how and information to predict and mitigate natural and man-made disasters.

The first civilian application of space-based remote sensing occurred in the 1960s with the development of civilian weather satellites, creating a revolution in meteorology, climate studies, weather warning and enabling greater comprehension of terrestrial weather and the forces affecting it. Meteorological services have since matured to a common ubiquitous utility generating widespread benefit. This was followed by the launch of environment monitoring satellites to provide environmental inputs for use by scientists, agriculturists, water resource managers, etc. The ability to undertake repetitive observation of Earth from space provided enormous data and information which spawned the growth of multi disciplinary sciences like Earth Systems Science (ESS) as well as industry and multinational endeavours aimed at applying the information for commercial gain and betterment. Because of its ability to provide up-to-date information quickly over a large area, space-borne imagery is increasingly being adopted by the mapping industry. Unlike airborne platforms, satellites provide regular, systematic and synoptic views of all areas of the globe with a consistent geometry and at a reasonable cost per square kilometre, thus enabling effective land-monitoring practices and repeatable cartographic analysis. Problems of access to remote and restricted areas are also overcome.

Imagery generated by space-based remote sensing is put to a variety of uses. Most imagery includes photographs generated by visible light, (a high-altitude replacement for aerial photography); infrared images that record heat and levels of radiation; and multi-spectral images that detect the presence chemicals, minerals, etc. on Earth. Primary application of this imagery revolves around

- Resource management and allocation: Management and distribution of resources like water, ocean resources, energy (oil and gas exploration), forests, agriculture.

- Disaster warning and mitigation: Warning and post-disaster mitigation in case of disasters like storms, floods, wildfires, volcanoes, landslides, earthquakes, etc.

- Community growth and welfare: Infrastructure development, urban town planning, land use practices, conservation, utilities, transportation and so forth.

- Environment assessment: Assessment of changes and effect on environment.

The terrestrial uses of remote sensing include land use/cover mapping, geologic and soil mapping, environmental and climatology assessment, landform identification, and applications for agriculture, forestry, rangeland, biomass, water resources, urban and regional planning, wetland mapping, wildlife ecology, archaeology, meteorology, etc. In addition, numerous miscellaneous applications ranging from crop health patterns, illegal oil discharges to unearthing mass graves as well as hidden narcotic plantations, gauging cocaine outputs,[14] etc. are in vogue and serve humanity. It has application not only of interest to agriculturists, environmentalists, oceanographers and the like, but to historians, archaeologists and palaeontologists. For example, remote-sensing satellites also helped unearth remains of ancient cities like Ubar (founded circa 2,800 B.C.E) in Oman, Sri Krishna's Dwarka in India, and so on.

International efforts to reach the benefits of Earth observation arrived following the initiation of the International Charter on Space and Major Disaster (ISMD) at the Third United Nations Conference on the Exploration and Peaceful Uses of Outer Space (UNISPACE III) in Vienna in 1999. The original members of the Charter were the ESA, CNES and the CSA, since joined by the Indian Space Research Organisation (ISRO), the United States National Oceanic and Atmospheric Administration (NOAA) and the Argentinean space agency CONAE. By 2000, the Charter had become operational and provided succour following landslides in Slovenia. In 2002, it covered up to seven major disasters worldwide. In 2003, it provided assistance following the earthquake in Bam, Iran and also provided satellite imagery of massive forest fires following invocation of the Charter by Portugal's Civil Protection Agency. As in September

2004, the Charter has gained several new signatories among the world's space agencies and has been activated more than thirty times. It is perhaps the single most dramatic example of how space technology can benefit and protect citizens of the globe.[15]

The benefits of Earth observation unlike those of satellite communications are not quantifiable by the amount of revenue generated as numerous applications are aimed at human welfare rather than commercial gain. Compared to the market for commercial communications satellites (or even for civil and military types), the commercial Earth imaging satellite market is small. Only 43 such spacecraft are forecast to be built and launched worldwide between 2001 and 2010. Market observers like the Teal Group believed that they would account for only about 3% of all satellites over the next 10 years, in terms of units. At an estimated worth of $3.62 billion, they will also make up only about 3% of the total value of all satellites.[16]

In addition to these numerous uses of navigation, position and location exist and are widely known and in practice. GPS receivers gain ubiquity with every passing day and are used for everything ranging from vehicular navigation, transport and resource management, search and rescue to tracking of parolees, pets and errant children, etc. Such applications have become part and parcel of everyday life and revolutionised lifestyles globally as never before. GIS and GPS applications have spawned and sustain an industry which experiences continuous growth and development.

Revolutionising Commerce

Growth in Revenue

The space industry has been growing progressively over the last four decades and continues to grow on a positive scale. The proliferation of the Internet, expanding demand for telecommunications in emerging markets, the need for Earth observation applications as well as acquisition of new military capabilities, growing concern for the environment, etc. present tremendous opportunities for growth in the satellite industry. World satellite industry revenues have grown pro-gressively, for example, from $38 billion in 1996 to $91 billion in 2003.[17] Sector-wise growth in the satellite industry is as indicated in Figure 2.

Satellite Industry Revenues[18]

Satellite Industry	1996	2003
Satellite Services	$15.8 bn	$55.9 bn
Satellite Launch Industry	$4.2 bn	$3.2 bn
Satellite Manufacturing	$8.3 bn	$9.8 bn
Ground Eqpt Manufacture	$9.7 bn	$22.1 bn

Figure 2

Commercial Impact of Space

Annual global expenditures on commercial space activities have exceeded government expenditures and the disparity is widening each year. In the US alone, the value of commercial space ventures has grown from approximately $59 billion in 1995 to well over $100 billion in 2000—a 70% increase in just five years, according to George Morgan, Executive Director of the Space and Wireless Business Centre at Virginia Polytechnic Institute and State University.[19] The Satellite Industry Association estimates that the worldwide commercial satellite business already represents a $44-billion industry, providing over 150,000 high-wage, high-tech jobs. Roughly half of those revenues and jobs are in the United States. Annual growth in this area was over 14% in 1997 and is projected to remain strong as the global demand for satellite services expands.[20] According to the Teal group, between 900 and 2,000 satellites are expected to be launched in the next decade. Translated, this is at least 90 satellites *per year,* and this figure does not include the hundreds or possibly thousands of nano-satellites that individual companies may also launch.[21]

Satellite Industry Trends

As per the release of a report by the International Space Business Council in August 2005, the following are the trends in the satellite industry:[22]

- World turnover generated from commercial services and government programmes reached $103 billion in 2004 and is forecast to exceed $158 billion in 2010.
- More than $18 billion is spent annually on the development of space systems.
- US Defence spending on space has grown from around $15 billion in 2000 to more than $22 billion today and is forecast to reach $28 billion by 2010.

- India and China have joined the US, Europe, Russia, and Japan as having fully independent capabilities.
- Satellite-to-consumer television has become a $40-billion worldwide market.
- The markets for satellite radio and GPS positioning and tracking are being validated with growth measured in billions.
- The successful launch of new satellite broadband services in the US and Canada beginning in 2005 could improve the market for commercial infrastructure.
- The development of a substantial space tourism market would have a positive but disruptive influence on the industry, though it is not likely to happen before 2010.

Small, Micro, Mini and Nano-Satellites

In addition to conventional satellites, technology has also enabled the development of small satellites which are set to revolutionise the use of satellites by making space access more affordable. Along with reduced development times, the inherent reduction of launch costs offered by the reduced size and mass of the spacecraft and their more manageable proportions, small satellites also offer an affordable option to develop and establish expertise in space technology and access requisite missions. Small satellites are not a solution for all types of mission, but offer the possibility of performing ambitious scientific experiments and applications as a complement to large missions. Their capabilities would gradually increase with improvements in electronic processors and sensors.

There is no universally accepted definition of a "small satellite". Normally, the generic term "small satellite" is used for a spacecraft of less than 1,000 kilograms. Below that limit, satellites over 100 kilograms are frequently called "minisatellites", between 10 and 100 kilograms "microsatellites" and below 10 kilograms "nanosatellites". The cost of developing and manufacturing a typical minisatellite is between US$5 million and 20 million, a microsatellite between US$2 million and 5 million and a nanosatellite could be below US$1 million.[22]

Other Uses

Apart from these, numerous other utilities and spin-offs accrue from space-related activities. These diverse applications range from advertising endeavours like Pizza Hut placing its logo on the fuselage of a 200-feet Proton rocket which launched the ISS in November 1999 to space tourism,[24] space funerals and burials undertaken by firms like Celestis. The firm enables individuals to have their cremated remains blasted into space. The cost depends on launch distances. A short launch into low-Earth orbit costs about $5,300, but if the remains need to be flown to the moon or launched into deep space the price rises to more than $12,000.

Less frivolous applications and uses like scientific study of the complexities and dynamics of the universe, microgravity experiments, utilisation of celestial resources, etc. are also in vogue and are devoted to furthering the cause of humankind. Exploration of the means of tapping space-based energy resources like the energy of the sun; helium resources on the moon, etc. is also being undertaken. It has been estimated that the moon has at least a million tons of He3. Theoretically, a ton of He3 can produce 10,000 megawatts of electricity for a year, enough for a city with a population of 10 million.

Contribution of Space to India's National Development

Church Promoting Science

Traditionally, relations between the Church and space science across the globe have been characterised by animosity. Religious intolerance of science at times led to the risk of being burnt at the stakes. However, the reverse applies in the case of India. The Catholic Church at Thumba in Thiruvananthapuram, Kerala not only helped in securing land for India's early space endeavours, but also agreed to vacate the St Mary Magdalene Church which fell within the area and actually offered the Church and its parsonage for use as laboratories and offices of India's space programme. These were the first laboratories and offices of the Indian space programme.[25] For a nation that began by transporting its rockets on bicycles and payloads on bullock carts four decades ago, its present levels of achievement are truly noteworthy. It has presently reached a stage wherein its space programme not only reckons with the best in the world and aspires to reach the moon, but also drives national development, welfare and commerce as no other single national programme in the world does.

Early Developments

The Indian space programme is focussed at national civil development rather than military needs, unlike in most cases globally. The programme evolved through three distinct phases to support national needs. The first was the initiation phase of the 1970s when the programme was extremely nascent. This was followed by the experimental phase of the 1980s, when scientific studies, end-to-end applications like vantage points of space for remote sensing, communications, etc. were studied and worked upon. These experiments gave rise to efforts like the *Bhaskara* for remote sensing, and APPLE[26] for communications. This was followed by the operational phase of the 1990s, wherein systems were built and established which could be routinely applied for developmental purposes. These applications were largely in the realm of communications, broadcasting, remote sensing, meteorology, etc. This period was characterised by an expansion of capabilities with systems being built, launched and put to civil application and uses. The increasing needs of national development were instrumental in speeding up the third phase. By the millenium, the system had matured to a level wherein it stared offsetting the expenditure incurred both by commercial gain as well as the benefits provided in terms of welfare and development. The last four decades since inception were witness to the launching of over 40-odd missions to promote the cause of welfare and national growth. Space technology, in the last four decades has firmly established its capability for socio–economic development and welfare in the country. Incremental growth in capabilities was a natural offshoot of expanding uses. Satellite footprints have graduated beyond national to regional levels. For example, satellites like *INSAT2E* have footprints covering South East Asia to mid-Europe. Remote-sensing capabilities have progressed from one kilometre resolution of the Bhaskara variety to 70 metres, from 70 to 23 metres, to five and on to one metre and even less.

Launch capabilities

Launch capabilities have also significantly improved as is evident in the successful development of Polar Satellite Launch Vehicle, PSLV, capable of putting 1,000–1,200 kg class satellite into an 820-km polar sun-synchronous orbit. PSLV is being offered to launch satellites of other countries. It launched two small satellites, one of Korea and another of

Germany along with India's IRS-P4 in May 1999. The Geo-synchronous Satellite Launch Vehicle (GSLV-D1) had its maiden successful flight on 18 April 2001 from Sriharikota, injecting the G-SAT 1 satellite into geo-synchronous transfer orbit. India has already carved a place for itself in the launch business. It has launched small satellites for Belgium, Germany and South Korea and has contracts to launch one for Singapore in 2006 and an ultraviolet telescope for Israel in 2006. In 2007, India's ambitious moon probe will carry scientific payloads from Europe and the United States.

Space Applications

The list of space applications furthering the cause of national development and welfare is significant. The intent is not to list or recount all of them, but to provide a brief overview with an end to comprehending the impact of space upon national development. India's outstanding use of space is in applications of remote sensing and communications. India has established the largest constellation of remote-sensing satellites in the civilian sector, the Indian Remote Sensing (IRS) system. These satellites provide valuable data on natural resources, not only domestically, but also to more than a dozen countries around the world. It also has one of the largest domestic communication satellite systems in the Asia–Pacific region, the Indian National Satellite (INSAT) system. The system comprises up to eight communications satellites operated by ISRO. These are now used by 35,000 commercial customers, all based in India. With more than 130 telephone exchanges in orbit, the Indian National Satellite (INSAT) system is one of the largest domestic communications systems in the Asia–Pacific region. Thanks to INSAT, 90 per cent of Indians can watch television. Daily weather forecasts and disaster warnings are possible. The INSAT is primarily utilised for telecommunication, television, radio networking, meteorology, search and rescue, etc. whereas the primary applications of data from Indian Remote Sensing Satellite is used for various applications of resources survey and management under the National Natural Resources Management System (NNRMS). Some of the applications are as under.[27]

- Pre harvest crop acreage and production estimation of major crops.
- Drought monitoring and assessment based on vegetation condition.
- Flood risk zone mapping and flood damage assessment.

- Hydro-geomorphological maps for locating underground water resources for drilling wells.
- Irrigation command area status monitoring.
- Snow-melt runoff estimates for planning water use in downstream projects.
- Land use and land cover mapping.
- Urban planning.
- Forest survey.
- Wetland mapping.
- Environmental impact analysis.
- Mineral Prospecting.
- Coastal studies.
- Integrated Mission for Sustainable Development for generating locale-specific prescriptions for integrated land and water resources development in 174 districts.

Impact on National Development

In addition to these, space impacts daily life and improves the quality of life in a variety of ways, some of which are recounted below. Initiatives like Village Resources Centres (VRCs) aimed at providing locale-specific information to the rural population by effectively combining the communication capacity of the INSAT and Earth observation capability of the IRS are changing the quality of life in the remote corners of the country. VRCs facilitate access to spatial information on important aspects like land use/land cover, soil and ground water prospects and enable farmers in taking important decisions on agriculture. Besides, VRCs also enable online interaction between the local farmers and agricultural scientists and provide critical information to fishermen like state of the sea and wave heights. Provision of information on many governmental schemes, farming system, action plans based on weather, community-specific advice on soil and water conservation, etc. are the other services.

In September 2004, ISRO achieved a world first with the GSLV-launched 2-tonne *EDUSAT,* the world's first satellite dedicated to providing support for educational projects. *EDUSAT* would initially link 5,000 schools and colleges in five states and eventually expand into a nationwide space-based educational service. This system is primarily meant for school, college and higher levels of education and to support non-formal level of education.

With regard to the telemedicine initiative, the system consists of customised medical software integrated with computer hardware along with medical diagnostic instruments connected to the commercial VSAT (Very Small Aperture Terminal) at each location. The medical record/history of the patient is sent to specialist doctors, who study and provide diagnosis and treatment during videoconference with the patient. As of 2004, the telemedicine network has been further expanded and now covers 100 hospitals, 78 remote/rural/district hospitals/health centres connected to 22 super speciality hospitals located in the major cities. More than 25,000 patients have so far been provided with teleconsultation and treatment. An impact study conducted on a thousand patients has revealed that there is a significant cost saving in the system since the patients avoid expenses towards travel, stay and treatment at hospitals in the cities.[28]

Conclusion

In addition to these numerous applications in agriculture, fisheries, health, water, etc., space has rendered enormous service and has more than justified the investment of the government. The level of contribution to national development can be gauged from the Indian Prime Minister's statement on the subject that the "Main key areas of our national effort are dependent on the optimum use of our space assets".[29] This not only commends the Indian space effort but also validates the enormous contribution of space and the specific community dedicated to reaching the benefits to the populace.

Notes and References

1. As on 31 December 2003. Ref Tamar A. Mehuron, Space Almanac, p. 49, *Air Force Magazine,* August 2004.

2. Signal Communications Orbit Relay Experiment.

3. Michel Andrien & Pierre–Alain Scheib, "Space: the forgotten frontier," *OECD Observer,* April 2003. Available at www.oecdobserver.org/news/fullstory.phd/aid/963/Space.

4. For a more complete brief, see Larry Valenciano, Group Director Intelsat

"Satellites—The Growing Demand and Reach," *Connect-World* at www.connect-world.com/Articles/Larry Valenciano.htm

5. Sean Moroney & Paul Hamilton, "Satellite and VSAT: Innovative uses for rural telephony and Internet development", AITEC, Africa, p. 1.

6. Ibid., p. 4.

7. For example, Shazda Ansari of Mumbai got married to Kasim Ansari from Jeddah in an Internet *Nikaah*. See *The Indian Express*, 2 December 2005, p. 3

8. Above figures provided in 1999 by *The Yankee Group*, a US marketing research. See Larry Valenciano's brief, "Satellites—the Growing Demand and Reach."

9. Quoting from survey conducted by the Internet and Mobile Association of India (IAMAI), See "54% more Net users in India," *The Times of India*, 9 December 2005, p. 15.

10. See *Wikipedia*, "Background History—Intelsat" at en.wikepedia.org/wiki/intelsat/htm

11. INELSAT was sold off to four private equity firms in August 2004 at a cost of $3.1 billion.

12. Quoting from speech of Jim Albaugh, President Space & Communication, The Boeing Company at "International Conference & Exhibition Keynote Address," Satellite 2000, Washington D.C. available at www.boeing.com/news/speeches/2000/albaugh000203.htm

13. Henry S. Kenyon, "Pentagon Boosts Telecomm Business into Higher Orbit," *Signal Magazine*, April 2004. Available at www.afcea.org/signal/articles/anmviewer.asp%

14. South America's Cocaine output rose by 2% in 2004 according to a UN report, see www.signonsandiego.com

15. Ref site of European Space Agency, "Civil protection assistance," at www.esa.int/esaEO/SEMARE3VQUD_environment_0.html

16. Marco Caceres, Teal Group, "Focus sharpens for imaging satellite market," *Aerospace America*, September 2000 at www.aiaa.org/aerospace/Article.cfm?issuetocid=8&ArchiveIssueID=5

17. Figures have been sourced from report of *Futron Corporation* "2003 Satellite Industry Statistics," June 2004.

18. Ibid.

19. Quoting from Shari Caudron, "The Vast Reach of Space Commerce," *Business Finance*, July 2001. Available at www.sharicaudron.com/articles/space.html

20. Quoting from remarks of Gary R. Bachula, acting under secretary for technology, US Deptt of Commerce, delivered on 23 September 1998. Available at www.technology.gov/Speeches/p_commsptr.htm

21. Shari Caudron, "The Vast Reach of Space Commerce," *Business Finance,* July 2001.

22. Ref report released by ISBC, "Space and Satellite Market Surpasses $103 bn to reach $158 bn by 2010" at www.californiaspaceauthority.org/html.

23. Ref *Background Paper-9,* "Small Satellites," Third UN Conference on the exploitation and Peaceful Uses of Outer Space, A/CONF.184/BP/9 dated 26 May 1998.

24. Space tourism is primarily being promoted by Russian agencies like Mir Corp, Russian Aviation and Space Agency, etc. Californian millionaire Dennis Tito undertook a week-long visit in space after paying $20 million to the Russian Aviation and Space Agency.

25. The church has presently been converted into a space museum.

26. APPLE stood for Ariane Passenger Payload Experiment.

27. Site of ISRO on space services. See www.isro.org/space_services.htm

28. For a more complete brief of the impact of space in revolutionising life in India, visit site of ISRO at www.isro.org/rep2005/SpaceApplications.htm

29. Prime Minister Manmohan Singh's address to scientists at Sriharikota on 21 September 2005. Ref site of PMO at http://pmindia.nic.in/speech/contest.asp?id=196.

Chapter 3

Merging Frontiers of Air and Space

Since the dawn of civilisation, human beings have always sought to acquire the "high-ground" for furtherance of their military aims and missions, as evidenced by the sense of cavalry, war elephants, aviation and now space. The inherent attributes afforded by elevation like reach, perspective, mobility, etc. made it a much sought-after tool of warfare. Thus, while controlling the high ground has been a rule of warfare ever since the dawn of civilisation, as warfare moved from the earth's surface into the air, the military advantages of control of the high ground became even more pronounced and air supremacy (or air superiority, at least) became a prerequisite for any successful military campaign. As in case of the early days of aviation, when the primary contribution of aviation to warfare was reconnaissance and observation, the space age also began as a supplement to aerial reconaissance with space-based reconnaissance as its main endeavour.

As a field for military operations, space literally never existed as a real possibility prior to the twentieth century. It was in World War II that the possibility of using space as an arena for military operations was first introduced. In a fundamental sense this use of space was confined to rockets like V-2 using the lower fringes of space as an uncontested pathway to targets. However, the use of rockets was conceived initially as an extension of long-range artillery. In the case of V-2, although it traversed through the fringes of space, it was not explicitly a space vehicle even though its flight to targets was uncontested, which led to perceptions on using space as a medium of transit for delivering munitions. However, these perceptions were tempered more by the technical difficulties of those times rather than any broad conceptualisation of space as an arena for future military operations. This is validated by the fact that the next successor of the V-2, the ICBM, did not fly successfully for almost a decade after World War II.

The actual impetus for military utilisation of space beyond just delivery of munitions arose following the failure of American high-altitude strategic reconnaissance aircraft like the U-2 to photograph and locate Soviet ICBM sites.[1] Following the shooting down of a U-2 by a Soviet Air Force surface-to-air missile and recovery of the pilot and the aircraft by the Soviets, the Americans had to suspend the U-2 programme and look towards safer and better reconnaissance options afforded by the higher realm of space.

Analogous to aviation in its early days, neither the Soviets nor the Americans had a comprehensive doctrine related to the potential uses of space, and reconnaissance was then deemed as the sole viable utility of space-based capabilities. However, as with the U-2, the Soviets threatened to shoot down US satellites and began developing ASAT (Anti SATellite) weapons. The Soviets developed several systems in the 1960s and tested them many times with varied though promising results. Not to be left behind, the Americans developed their first ASAT programme, the Air Force's SAINT (Satellite Interceptor) programme. Thus within three years of *Sputnik-1* ushering in the space age in October 1957, effective military utilisation of space had begun and while the prospects of using military space seemed infinite, reality proved more prosaic. It took even a technologically advanced and economically wealthy nation like the US, the best of over four decades to begin a weaponisation programme in space.

Nevertheless, there is no contesting the fact that space has been militarised for over 40 years. Reconnaissance, surveillance, warning, communications, weather, and navigation satellites were designed and deployed to enhance conventional military prowess and serve national security needs.

Analogous to air power, which early in the 20th century emerged to support and later protect surface forces and whose role expanded to that of a separate instrument of warfare meant to protect national interests as also ensuring freedom of action of surface forces, space power now holds the promise to do the same and even more in the 21st century. As we move ahead in the 21st century it is inevitable that space will become another medium of warfare besides being an important economic centre of gravity. Closer home, within Asia, there exists clear evidence that the phenomenon of space militarisation is no longer confined to industrialised nations like the US. Countries like China, Japan, Israel, etc. already have military satellites in orbit and are pursuing military space capabilities. In fact, following the collapse of the Soviet Union, in terms of military space capabilities, China is closing in as the next closed rival to the US.

Systems related to national security have dominated space, but this dominance is eroding rapidly. The increasing number of countries and consortia turning to space to provide and receive services, and to generate wealth, will force nations to adapt to this emerging environment.[2] Achieving advantage in space power before and during conflicts will be critical to success on the battlefield.

Frontiers of the Vertical Dimension

In view of the above, an exposition of the two mediums of the vertical environment, air and space which enable the prosecution of decisive military operations and which in recent years are becoming demonstrably potent and pervasive as a singular operational entity would be in order to fully comprehend the attributes as well as challenges afforded by exploitation of these two mediums of the vertical dimension.

Technology-intensive Frontiers

It is common knowledge that man's only natural environment is land. Apart from land, all the other environments of sea, air, and space require technology of some sort to support human activities. Air and space demand even more technological support to even enable human presence. As man attempts to reach higher, the demands on technology increase correspondingly. Thus as technology evolves, the doors to greater and optimal exploitation correspondingly open further and wider. Both air and space require enormous technological investment to sustain, support and exploit the distinctive characteristics offered by the unique environments. It took mankind many centuries to cultivate the environment of air so as to first tolerate human presence and then allow heavier than air activity in 1903, and ever since technology has been continuously opening up the vertical dimension. However, distinct environments have distinct physical characteristics and consequently demand different technologies for optimal exploitation. At this stage it is relevant to note that atmosphere, where the laws of aerodynamics apply, has no abrupt cut-off. It slowly becomes thinner and fades away into the emptiness of space. There is no definitive boundary between the atmosphere and outer space.[3] Three-quarters of the atmosphere's mass is within 11 km of the planetary surface. The highest altitude obtainable by air-breathing aircrafts is about 28 miles. In general parlance, the Karman line, at 62 miles (100 km), is also frequently used as the boundary between atmosphere and space because at around that altitude, the parameters required for aerodynamic flight are no longer available. Hence terrestrial-based forces generally operate below an altitude of roughly 62 miles whereas space-based forces operate above this altitude where the effects of lift and drag are negligible. Sustained flight thereafter is not impossible, but is obtained with tremendous expenditure of fuel. After that the next lowest altitude for sustainable satellite orbit is at around 93 miles. Thus, broadly three different kinds of technologies apply and cultivate the usable infinite vertical expanse above the surface of the earth. This presents varied challenges (as well as opportunities) and allows varied levels of flexibility of the environment. While aerodynamic technology has sufficiently matured in the atmosphere during the past century, technology is yet to allow the same or proximate levels of utility for the layers beyond the atmosphere ever since it opened up in the past four and a half decades.

Before space power became a military consideration, air power was (and still is) the primary instrument for exploiting the vertical expanse above the earth in support of national objectives. Technological limitations constrained the exploitation of the vertical dimension up to the atmosphere; however, with the launch of the *Sputnik,* the vertical realm beyond the atmosphere also opened up both for commercial and military exploitation. Technological advancements ever since have fused the mediums of air and space to the extent that apart from allowing the migration of certain aerial capabilities like observation, reconnaissance, navigation, etc. further up into space, it has integrated air and space-borne capabilities into terrestrial war fighting so much so that precision war fighting has now become the most sought-after norm rather than the exception. It was in the crucible of the 1991 Gulf War that the immense synergistic potential of air and space was first fully appreciated, recognized and demonstrated; ever since, nations across the world ranging from space super-powers like the US to those with nascent capabilities like Israel, South Korea, etc. are overwhelmingly focussed on integrating the attributes and advantages afforded by the mediums of air and space into their national war-fighting capabilities.

Overview of Attributes and Differences

In spite of modern technology and its potential to allow greater exploitation of the vertical medium, a critical comparison of the medium also needs to be undertaken in order to identify the similarities, differences and challenges so as to harness the parallel capabilities, mitigate the limitations and accept the challenges to unrestricted integration. The following Figure 3.1 briefly highlights the environmental attributes and differences of the two mediums of air and space.

Environmental Attributes and Differences

	Air	Space
Attributes	● Elevation ● Freedom of movement ● Technologically-cultivated environment	● Elevation ● Freedom of movement ● Technologically-uncultivated environment
Differences	● Aerodynamics ● Limited mission duration ● Transitory operations ● Legally restricted to political boundaries	● Astrodynamics ● Longer mission duration ● Positional operations ● No political boundaries

Figure 3.1

A causal examination of Figure 3.1 would reveal that apart from the physical differences lent by the laws of astrodynamics and aerodynamics, the other differences are not extremely pronounced. As technology and interoperability of the environments advance, the differences would not inhibit the pursuit of war-fighting concepts from the vertical dimension, but rather complement capabilities and compensate for deficiencies. For example, the limited mission durations of aerial reconnaissance, navigational platforms, etc. could be substituted by satellites with longer mission durations; similarly, the inadequacies of transitory operations could be compensated by positional operations and vice versa. Lastly, the question of political boundaries is more of a man-made restriction and not a difference attributable to the basic character of air. Nevertheless, absence of political boundaries in space in no way inhibits or retards exploitation of the vertical dimension for military purposes; on the contrary, it compensates for the restrictions imposed on aerial platforms. Additionally, as in case of aerial platforms, greater and better capability to exploit the environmental character would decide war-winning capabilities amongst rival nations. Just as nations with better air power capabilities would have an edge in warfare, those harnessing space better would wrest the advantage in space and consequently on terrestrial battlefields, thus seminally influencing the outcome of the battle.

Space technology as it develops shall facilitate greater exploitation of the entire vertical dimension. Technologically, different means of access and operation are required on account of the distinct environments; however, technological differences in the operating platforms or the means of access are not sufficient to necessitate demarcation of distinct frontiers or distinctly separate force structures. For example, submarines and surface ships have many technological and operational differences, but none compelling enough to deny or overwhelm the basic fact that both operate at sea and hence share the unifying properties of the same environment.[4] The same is the case with Infantry and Armour as also the entire spectrum of platforms in the vertical dimension ranging from hovercrafts, UAVs, drones, aircrafts, spacecrafts to ICBMs, etc. operating from Earth's surface, the exosphere to the stratosphere and beyond into low Earth orbits and farther. The important characteristic that air and space platforms share is that both operate above Earth's surface. Elevation here is a characteristic that changes only in amount not kind as one ascends from air into space.

This also validates the argument that from a military, as opposed to an engineering perspective, the environment must be considered as an indivisible whole rather than as two disparate frontiers. While it is obvious that environmental differences do exist between air and space, the same are not profound enough to contest the operational attributes necessitating conjoint utilisation of resources from the vertical dimension into terrestrial military operations in support of national security objectives.

Examining the Challenges to Integration

The nature of differences limiting conjoint operations are largely as indicated in Figure 3.2. The challenges are largely of a temporary nature and it would only be a matter of time before technology provides the necessary solutions for synergistic application of "aerospace power" in pursuit of national security objectives.

Space-based perspective inherently offers greater vertical depth and a broader field of view enabling near-instantaneous utility at a global

Differences Limiting Conjoint Operations

Medium	Environmental Attributes	Employment Differences	Nature/Character of Differences
Air	Perspective	Regional/Theatre	Permanent
Space	Perspective	Global	
Air	Range	Regional/global (refuel)[5]	Semi-Permanent
Space	Range	Global	
Air	Manoeuvrability	3-D, unlimited	Temporary
Space	Manoeuvrability	Limited, (orbital)	
Air	Responsiveness	Active & Passive	Temporary
Space	Responsiveness	Passive	
Air	Versatility	Unlimited	Temporary
Space	Versatility	Limited	
Air	Flexibility	Unlimited	Temporary
Space	Flexibility	Limited	
Air	Force application	Unlimited	Temporary
Space	Force application	Extremely limited	
Air	Accessibility	Easier (relatively)	Temporary
Space	Accessibility	Difficult	

Figure 3.2

level as opposed to air wherein the same would be limited to the area covered by attaining particular altitudes This difference is of a permanent nature and would always remain so. However, with regard to range, the physical limitation of fuel inhibiting global ranges was overcome the day a non-stop, non-refuelled *Voyager* aircraft flew around the globe in 1986. It conclusively illustrated the fact that both the mediums of the vertical dimension, i.e. air and space enabled global ranges on account of the common environmental characteristic of ubiquity. However, the levels of flexibility, versatility, manoeuvrability and accessibility inherent in airborne platforms are not presently matched by space platforms. Nevertheless, this is not a permanent insurmountable difference and advancing technology is already in the process of overcoming these challenges considering that multi-tasking of satellites is already under way, alternate fuel options to increase menoeuvrability of satellites are being explored and finally space accessibility today is much easier and cheaper than it was in the last decade. In addition, differences like military force application from space are restricted more by prevailing legalities rather than technology or the environment and a casual glimpse of the political horizon in space would indicate that this difference would soon be diluted or even dispensed with in the near future.

Thus, an operational or functional approach as opposed to a biological, geographical or technological approach to the thinking about the synergistic application of air and space power in the vertical dimension makes the differences between orbital (space-based) and atmospheric (aerial) operations appear almost irrelevant. Technological evolution of air-breathing platforms like aeroplanes into hyper planes designed to exploit both the aerial and space environments are on the brink of opening up new paradigms for gainful exploitation of the vertical dimension. The imminent development of aerospace planes to exploit the full spectrum of air and space only serves to illustrate the continuity of air and space in the vertical medium. This continuity is further evidenced by the fact that conceptually many of the same military activities can be performed in air and space, even though different platforms and different methods are required to perform them.

Examining the Wisdom of Conjoint Exploitation

Since environmental attributes form the basis of determination of military force employment, these are dealt with first. The advantages of operating in the vertical dimension endow many military capabilities unique in their application and employment, thus making it a most versatile component of military power. The vertical dimension has two related attributes that distinguish it from the horizontal dimension. One, it has only one distinct boundary—the earth's surface; no lateral boundaries or geographical obstructions restrict movement within it, thereby endowing upon it the important quality of freedom of movement and its attendant advantages. Two, the vertical expanse extends from the earth's surface upwards and outwards to infinity, thereby endowing upon it the next important attributes of elevation and all-encompassing pervasiveness along with their attendant advantages. These distinguishing attributes afford a number of related advantages. From their elevated positions, air and space platforms have a broader perspective and, with freedom from surface constraints, they can travel faster, go farther and move through a broader variety of motions than surface forces. The primary advantages accrued on account of its unique attributes are

Perspective

● Elevation provides greater vertical depth and a broader field of view than can be obtained by observers, on the surface. This common attribute of air and space systems is one of the reasons which led to early systems in both cases being used for observation and reconnaissance purposes. Space forces, apart from complementing and compensating for aerial observation deficiencies, by virtue of being more elevated than aerial systems also provide a safer and better means of observation depending on the prevalent circumstances.

Range

● While horizontal ranges for both platforms are literally infinite, the vertical range of air breathing systems is generally considered to be limited up to 50–60 miles,[6] beyond this space complements and possesses the requisite attributes for compensating for such deficiencies.

Manoeuvrability

● Elevation and freedom of movement endow platforms operating in the vertical dimension with three-dimensional manoeuvrability as compared to surface forces which can only move on the earth's surface and are dependent on terrestrial features for movement. The mission requirements of platforms in the vertical medium determine their degree of manoeuvrability. For example, while fighter aircraft have exceptional manoeuvrability and agility on account of their mission requirements, geostationary satellites need just about enough manoeuvrability to maintain position over particular spots over the earth's surface. In either case, the point is that the environment permits the platform to accomplish its mission at high speeds, slow speeds or even stationary positioning or in requisite axes on account of the related factors of elevation as well as freedom of movement.

Responsiveness

● The two foregoing factors also endow platforms and systems operating in the vertical dimension with rapid response capabilities to accomplish assigned tasks rapidly in addition to reacting to situations. Apart from the immense force application, combat support and other rapid response capabilities of aerial platforms, space systems provide potent response options and immediate access in areas of interest with regard to surveillance, intelligence, navigation, positioning, early warning, etc. While rapid response options of space systems with regard to force application are presently limited by technological and legal factors, the potential for their use is immense.

Pervasiveness

- Air and space cover the entire globe, hence technically there are no natural lateral boundaries restricting movement. Thus while air forces can go wherever there is air, space forces go even further and do not even need air as a supporting medium. Nevertheless, while no physical obstacles obstruct aerial flights, legally speaking, nations do regard their air-space as sovereign and can deny over-flight rights even in peace time. Space's all-encompassing invasiveness, however, compensates for this because no political boundaries have been established in space and free access to the globe is available from space.

Thus the aforementioned advantages afforded by a conjoint application of air and space systems as opposed to singular hermetic applications would enable optimum exploitation of the vertical dimension in support of operational objectives. A combination of both the attributes of the air and space environment would enable better exploitation and would have a cyclic effect by providing additional related attributes like flexibility, versatility, speed, etc. Indeed, the Gulf War only served to validate the view that the overwhelming pervasiveness of the vertical dimension in all aspects of modern warfare is no longer in question.

The advantages afforded by the vertical dimension along with the unique capabilities of air and space power produce a wide variety of options and opportunities for accomplishing national security objectives. The above-mentioned combinations provide the foundation for its employability and the combination needs to be exploited as a single entity for ensuring weight of effect rather than dispersing it into the proverbial "penny-packets".

Distinctive Challenges of the Horizontal and Vertical Dimension

Environmental differences do exist between air and space. However, unlike land and sea, no distinct boundary separates the two mediums of air and space in the vertical dimension. A transition zone does exist, but again this does not separate the operational employment of forces of air and space as a singular entity operating in the vertical dimension, whereas in the case of land and sea, a natural and distinct boundary exists in addition to a transition zone. Land and sea power influence each other in the transition zone, but beyond land and sea forces generally fight independent of each other. However, the all-pervasive vertical envelopment afforded by aerial and space capabilities conversely affects both land and sea power from the vertical dimension.

Thus despite physical differences, air and space forces operating together offer a unique and potentially decisive synergistic effect from a vertical dimension as opposed to a horizontal dimension where distinct boundaries preclude the application of power as a single, seamless entity. This is not to be construed as an argument against joint application of force, but as a mature acceptance of the challenges posed by distinct boundaries as one moves from land to shore and vice versa in a horizontal dimension as opposed to the lack of such stringent overriding factors in the vertical dimensions. Hence there exists a consequent need to look for options aimed at mitigating the challenges and plugging such inadequacies through optimal exploitation of the vertical dimensions of air and space.

The vertical dimension comprising air and space has a unifying effect on the conduct of warfare since it envelopes the entire globe. For example, land and sea forces operated largely independent of each other till the advent of air power which enabled coordinated application of force. Inclusion of the space paradigm into the vertical dimension would only enable greater coordination of the military forces of land, sea and air to produce a more effective combat team. Instead of getting tied down by the fact that air and space environments dictate exploitation of different laws of physics for operation, it needs to be understood that these physical differences do not overwhelm the commonality that the operational characteristics of air and space afford, like elevation, freedom from geographical constraints, perspective, range, etc. The physical differences, contrarily, only provide a means of better optimal exploitation of the vertical dimension encouraging complementary and conjoint exploitation rather than hermetic applications and exploitation as individual, independent mediums. The entire characteristics afforded by the vertical spectrum need to be exploited and integrated into terrestrial military machinery to obtain decisive war-winning capabilities. The eminent logic of this strategy was evident in the recent Gulf War, when air and space power complemented each other operationally through command and control of air missions, acquisition and targeting, demonstrating that a synergistic solution to operational problems could be obtained by effectively harnessing and integrating air and space capabilities of the vertical medium into terrestrial operations for a pronounced weight of effect.

Thus space, instead of being a frontier provides an extended dimension to air power in the vertical dimension by enabling a virtual bridge to capabilities like reconnaissance, navigation, early warning, etc. which in turn provide the capabilities afforded by the much sought-after "high ground" of observation, perspective, force application, etc. In view of the foregoing, air and space power are apparently complementary capabilities essential for prosecuting successful operations. The future would hence comprise a mix of air and space forces executing vertical dimension operational concepts in support of national security objectives.

Doctrinal Analogy

As in the case of air power, space power is also capital and technology intensive, hence the principle that for optimal exploitation, the focus should be less on the "bean-count" (number of assets) and more on the correct rationale and doctrine for employment of such scarce and capital intensive assets applies in both cases. As with any evolving military field, intense debate over doctrine and correct employability of space can be expected. Like the emergence of sea and air power, space power has both similarities and differences with other forms of military and national power. However, focussing on similarities and differences for advocating ownership of assets or for that matter, focussing on the technical characteristics and numerical count at the cost of optimal doctrinal awareness runs the risk of finally ending up with possession of assets which are not gainfully employed in field.[7] In order to discount the possibility of employing space assets according to ill-suited doctrines and conventions, the endeavour should be to undertake a thoughtful consideration of the similarities, differences, characteristics, etc. of space assets for their optimum exploitation rather than simply basing doctrine on the general attributes of the space environment or blindly adapting doctrines of vastly advanced nations like the US.

The prevailing doctrines, missions and roles and space build upon prevailing theories of air power. Air power theory relied heavily on attributes of the maritime environment and naval experiences. Even terminology and rules of traffic and aviation are basically derived from the maritime environment. The sea offered less friction and greater maneuverability than land where terrain was a limiting factor. Air however, transcends both barriers of friction and manoeuvrability because it applies to all three dimensions. Space is also similar and applies to all three dimensions except that lateral movement is limited (because it needs more power and fuel) though it compensates for the same by higher speeds and altitude capabilities. At sea, sub-surface offers the third dimension of depth, though with severe limitations. Air and space have lesser differences than naval surface and sub-surface environments, but no one treats them differently.

Thus most space employment theories are premised on the contention that while space is a medium by itself, the distinguishing characteristics of air power like elevation, freedom of movement, perspective, etc. are common to both mediums on account of the principle of vertical envelopment and the common characteristics offered by the "high ground". Thus, space power employability builds upon the four basic employment principles of air power, viz. control of the environment, application of combat power, enhancement of combat power and sustaining combat power. It would be significant to note here that both the super-powers of the FSU and the US were broadly in common conformity as regards the environmental doctrine needed for employment of military forces within the unique environment of space; yet, their military space organisational doctrines were tempered by their distinct national requirements.[8]

Necessity of Conjoint Defence

Apart from an offensive paradigm, even when viewed from a defensive aspect, air and space are complementary to each other considering that both as complementary components of an air/aerspace defence system compensate for each others' deficiencies in countering threats of weapon systems like ICBM's which transit and manoeuvre through both the mediums of air and space.[9] The Soviets (as well as present-day Russians) strongly allude to this view and regard space as an extension of the aerial medium; a fundamental strategic operating medium, which, for doctrinal and strategic purposes is an extension of the medium of air. Thus, Soviet "air defence" contiguously extends vertically to include both air and space and consequently extends to include conventional air defence, ballistic missile defence as well as space defence. This finds reflection in the evolution of the Soviet Air Defence machinery (*ProtivoVozdushnaya Operatsiya*) PVO to air and space defence (*ProtivoKosnicheskaya Oborona*) PKO whose primary mission is to repel attacks emanating from space. The Soviet concept hence considered space as a part of a theatre of military operations or Teatr Voyennikh Deystyii (TVD) and their objective of space superiority was integral to, as well as an extension of, the overall objective of air superiority.[10]

The Counter-view

Akin to the vibrant dynamism witnessed in evolving doctrines, concepts and strategies for enabling optimal utilisation and exploitation of the aerial environment to afford war-winning capabilities, space is witnessing intense debate and discussion revolving around the basic premise of enabling optimal exploitability of the medium for furthering national objectives. Unfortunately, following the collapse of the FSU and the enormously nascent capabilities (in comparative terms of the US) of the next closest rival, the Chinese, a preponderance of American thought on the subject is in vogue. The rich divergence of views, counter-views and experience which shaped contemporary theories of air power is conspicuous in its absence considering the global disinterest of most nations, the depleting interests of the Russians on the subject, the linguistic complications as well as the vast capability gap of the Chinese. The point is that, unlike the diverse experiences of the Germans, the British, the Soviets, the Americans, Japanese and Koreans which continuously drove doctrine and practice of air power for optimal exploitation, space power theories draw on a limited reservoir of US and FSU experiences and following the collapse of the FSU, the lessons drawn are almost entirely American in character.

Nevertheless, the fact exists that the American experience is not one of failed or sub-optimal exploitation of the medium. On the contrary, the facts indicate that the US experience in space exploitation has been growing from strength to strength with every passing conflict. The validity of their doctrinal premise of conjoint exploitation of the aerospace medium as well as of that of Air Force stewardship in coordinating and integrating space-based capabilities to produce war-winning effects is incessantly evident and getting better with every conflict they have been involved in since the Vietnam War.

However, in order to allay concerns of intellectual persuasion driven by the plethora of US literature on the subject, an examination of the counter-view on the subject is in order and the same is undertaken. While lengthy treatises have been written on the subject, most counter-views revolve around the following aspects.

Disparate and Not Conjoint Mediums of Air and Space

This argument relates to the vertical dimension constituting two disparate and not continuous or conjoint mediums of air and space. Some authors have misinterpreted contentions promoting conjoint exploitation, like "there is no division…between air and space. Air and Space are an indivisible field of operations"[11] to point out that the two mediums of air and space of the vertical dimensions are on the contrary distinctly separate, governed by distinct laws of Physics like Aerodynamics and Astrodynamics. Hence the logic that the vertical expanse above Earth is a continuum of the third dimension is invalid.

However, the foregoing contention is based on Physics and not operational military art or the purpose of using the medium of space in the vertical dimension better for military purpose which is the subject of debate. Further, apart from the logic of conjoint exploitation, doctrinal analogy, etc. dwelt upon in detail earlier, the fact exists that the contention has been that while physical differences do exist, they are not profound enough to contest the conjoint operational exploitability of the dimension for producing war-winning effects. Rather the attempt is to harness the advantages and mitigate the disadvantages. The physical differences are to be leveraged to produce war-winning attributes and the logical soundness of the same was apparently evident in the Gulf War.

Semantic Confusion over the Term "Aerospace"

Semantic confusion and debate swirls around this term ever since the first time General Thomas D. White used it on 16 May 1958.[12] By combining the words air and space and describing the earth's atmosphere and the regions beyond it as one operational medium, the Air Force created a term that symbolised a new meaning. The term 'aerospace' focuses on treating both mediums as one instead of emphasising differences. The term has a history of entering and exiting USAF lexicon with due deference to doctrinal, political and developmental needs. For example, in the pioneer days since the Air Force argued that it should seek to control and apply force from space just as from the air, the aerospace concept inevitably came into conflict with the Eisenhower administration's "space for peaceful purposes" policy.[13] That administration saw the aerospace concept (and any other discussion of overtly military activity in space) as antithetical to its secret but highest-priority space policy as established by the National Security Council Resolution 5520 in May 1955.[14] Until the 1980s, the Air Force simply substituted the word aerospace for air and for the greater part of the 1990s the Air Force abandoned aerospace both conceptually and semantically. Air Force Chief of Staff General Merrill McPeak emphasised the importance of space assets in enhancing the combat effectiveness of coalition forces during the Gulf War by labeling the conflict as "the first space war" and then changed the Air Force mission statement in June 1992 by adding the words 'air' and 'space'. Shortly thereafter, in its Global Engagement vision statement of November 1996, the Air Force stated that "we are now transitioning from an air force to an air and space force on an evolutionary path to a space and air force".[15] By 1999, Chief of Staff, General Ryan had indicated that he wanted to go back to the term 'aerospace'[16] and later the term was again back in vogue and continues to this day. Nevertheless, the confusion is largely confined to the semantic and definitional levels and evidently has in no way inhibited actual American capability to prosecute military operations using the medium of space. In fact, most critics are in broad agreement about the conceptual connotations of the term and prominent analysts on the subject aver that "conceptually, the roots of the aerospace concept are closely associated with airpower theory and run quite deep",[17] however they opine that while the idea of Aerospace may have been forward looking in 1958 it need not stagnate at that but needs to evolve further.

Evolving beyond Aerospace to Space forces

Most criticism of the Aerospace concept and Air Force stewardship is centred around the premise that just as air power evolved beyond ancillary support to ground forces into a full-fledged instrument of warfare, the time has come for space forces to undergo a similar evolutionary process and evolve beyond the Air Force into a separate instrument of warfare that would harness not only the capabilities but also the strengths of the environment of space.[18] This in effect means that instead of allowing complacency to set in, the environment needs to be exploited beyond providing support to military elements to actual force application and control of the environment.

Thus the counterview basically does not question the concept of continuity of the medium, but explores better force options for control of the environment (space superiority/space dominance) and for military force application from the environment. The general grouse is that the Air Force needs to steward space beyond force enhancement and support of space, notwithstanding its superb performance in successfully accomplishing force enhancement missions.[19] This could be the reason why environmental differences like manoeuvrability, gravity, political boundaries, etc. have assumed mammoth significance (e.g. a mission-oriented "aerospace craft" would need to be sufficiently manoeuvrable for multiple mission accomplishment and also have to re-enter its own political airspace). The political compulsions inherent in such an endeavour obviously demand a cultivation of opinion both within and outside the US. Thus, enormous scares of a space "Pearl Harbour", threats to commercial assets, etc. are bandied about along with the accusation that the Air Forces are under-utilising space and that analogous to control of the sea, control of space is essential to protect American assets in space. This strategy is evidently paying off considering that the US withdrew from the ABM treaty in June 2003 without much of a domestic or international uproar and has now embarked on space weaponisation programmes like the "Transformational Flight Plan" which is aimed at a whole range of products for space warfare ranging from air-launched ASAT missiles to LASERS, Hypervelocity rod bundles, etc.[20] apart from missile defence.

Doctrinal Pitfalls

As a natural corollary of this evolution, many existing doctrines and concepts are being questioned. The overwhelming opinion on the matter is that up to the stage of support of conventional military forces or missions of force-enhancement, the prevailing doctrines and concepts have served well. The need is to evolve further beyond using space capabilities and to harness the strengths afforded by the unique characteristics of the environment.[21] This rationale drives the quest for a better doctrine to carry the strengths beyond support to actual military force applicability from the realm of space on to Earth as well as for preserving extant capabilities. In such an endeavour control of space would be a pre-requisite, hence the emphasis on space control.

Budgetary and Organisational Concerns

The Air Force in the US has, beyond any shade of doubt been the pioneer as well as the steward in furthering its national military space capabilities. Nevertheless, questions have been raised regarding its stewardship, not in terms of performance but in terms of budgetary allocations which are mostly being spent on space rather than aerial assets.[22] It would also be interesting to note that even though the preponderance of military space assets are funded, owned, and operated by the US Air Force—90% of the DoD space people, 85% of the DoD space budget, 86% of the on-orbit DoD assets, and 90+% of the DoD space infrastructure[23]—the Air Force uses space as a critical enabler for the entire joint force. The Air Force does not consider space as a region or AOR to be owned as land is currently divided. Rather, it considers space like air in that it is a physical environment where functional missions are conducted to achieve military objectives.

The American experience reveals that while its Air Force space command was formed in 1982, its Naval command the next year, a Space command in 1985 and an Army space command in 1988, whereas the Air Force Space Command gradually grew in responsibility and resources, others mostly stagnated or maintained status quo. While initially its mission was confined to operating missile-warning satellites and sensors, and conducting space-surveillance activities, in 1985 satellite command-and-control responsibilities were conferred upon it. In 1990 the space-launch function, as well as the responsibility for associated launch facilities and down-range tracking sites, was transferred to Air Force Space Command from Air Force Systems Command. Though the other service commands and a unified command responsible for war fighting with space forces have existed for over 18 years, responsibility and stewardship of space forces has been an Air Force preserve for 23 years. The Air Force thus has served and continues to serve as the longest and best steward in furthering its national space capabilities though it never sought stewardship for itself, nor was it assigned the role of space-steward, stewardship was automatically devolved on to it. The fact that stewardship automatically devolved and continues with the Air Force till date, and that not only did it live up to its military commitments but improved its ability to exploit and integrate space to produce war-winning effects, conclusively validates the aforementioned doctrinal premises and logic of conjoint exploitation. This is precisely the reason why the US is set to combine all of the forces that the Air Force provides to USSTRATCOM—intercontinental ballistic missiles; space forces; information operations; intelligence, surveillance, and reconnaissance (ISR); and global strike—into a single component, which will be called Air Forces Strategic Command (AFSTRAT).[24]

Broad Inferences

In view of the foregoing, the following are apparent:

- While both the dimensions of air and space have distinct physical characteristics, they demand conjoint exploitation for harnessing the strengths afforded by distinct environments as well as mitigating the deficiencies for enabling optimal military utility in support of national security objectives.

- As technology advances, the levels of conjoint exploitability would correspondingly increase thereby reducing the challenges to unhindered exploitation of the vertical dimension.

- The overwhelming emphasis of nations has been on fulfilling their "Force Enhancement" missions and it has taken even an advanced nation like the US, the better of over four decades to undertake space control missions.

- Space provides support to all conventional military forces and there is no disputing the fact that space support would enhance the capabilities of every service's operations; however, what distinguishes the Air Force beyond enhancement of capabilities is the fact that subsequently some of its roles and missions would migrate into space as evidenced by the migration of high-level reconnissance from aerial to space-based platforms right from the beginning of the space-age to the present wherein many missions like navigation, meteorology, etc. have already migrated into space.

- Doctrinally, both the FSU (Former Soviet Union) and the US shared the same concepts of space power employment which are built upon concepts of air power. Also, in the absence of any countervailing theories suggesting better military space power employment, the above-mentioned concepts have taken root, are in vogue and would prevail for the present. Prevailing 'aerospace' theories and concepts are the most suitable for providing support to conventional military forces and integrating space support to terrestrial operations. In the absence of any better countervailing theory the same would prevail for the present. Nevertheless, the quest for better concepts aimed at taking space capabilities and strengths beyond support and integration is already on and might come up in the future.

- At the same time the technology is yet immature to allow full-scale migration from support and integration to actual combat or force application roles, and so is the evolution of distinct space forces to harness the strengths of the unique environment of space. Hence the aerospace concept for the present would continue to prevail, and and when the technology matures, the Air Force would continue to endure while space forces would emerge.

- It is common knowledge that environmental factors do not inhibit support amongst the organic forces, rather they serve to mitigate inadequacies and harness strengths. Stretching the analogy further, though space has unique environmental characteristics which would necessitate organic forces for combat in the environment, the same would not be the case for obtaining support from the environment of space and enabling integration of space capabilities into terrestrial war-fighting. As and when the technological, political and legislation develops (or permits) space forces could develop mere support to independent combat in support of military objectives just as the Air Force matured beyond surveillance and long-range artillery functions in support of ground forces. But till then, considering the vast similarities in terms of environmental characteristics, operational attributes, doctrinal applications, the experiences of nations like the US it would be obvious that the onerous responsibility of harnessing space for military capabilities naturally develops on the Air Force.

Conclusion

From the foregoing it is apparent that while physical differences do exist, nothing strongly inhibits the conjoint operational exploitation of the entire vertical expanse above the earth for producing terrestrial war-winning effects in support of national security objectives. The physical characteristics demand conjoint application to mitigate the environmental deficiencies as well as to harness the strengths. In view of the validation of prevailing premises on employment of "aerospace power" in the previous Gulf War as opposed to hermetic application of air and space power for delivering intense nuanced weight of effect and obtaining military objectives, the best possible course of action is conclusively clear.

Notes and References

1. Walter A. McDougall, "The heavens and the earth: a political history of the space age", New York: Basic Books Inc. p. 219.

2. Global trends indicate that space is increasingly becoming an economic centre of gravity, the loss or degradation of which would cripple commerce, finance, communications and numerous other activities, e.g. the scale of the worldwide value of commercial space, which was estimated to be $11.5 billion in 1991, increased to $57 billion by 2001. Assets of such high values obviously attract potential threats of many dimensions. Clashes of interest for optimal utilisation of the new medium would hence be inevitable.

3. For legal purposes, most major space powers generally accept "the lowest perigee attained by orbiting space vehicles as the present lower boundary of outer space", but this standard is not universal. Certain nations like Australia have enacted legislation demarching 100 km as the altitude beyond which domestic space legislation would mandatorily apply. Even though precise delineation between the two environments proves impossible, their physical differences remain significant. The space environment is largely a vaccum characterised by high-energy particles, fluctuating magnetic fields, and the presence of meteoroids and micro meteoroids. The motion of bodies in orbit

closely follows the laws of celestial mechanics, a much different system of knowledge than the laws of aerodynamics governing the flight of aircraft. Aircraft operate in the much more benign government of Earth's atmosphere, characterised by moisture, wind, precipitation, pressure, etc.

4. It would be pertinent here to quote an argument from the *USAF Basic Aerospace Doctrine, AFM1-1,* Vol. 2, p. 67 that "some people have seized on the differences in air and space technologies to argue that space constitutes a separate environment from the air and that space requires development of a separate force to exploit it just as land, sea and air environments require separate forces. This argument is equivalent to saying that submarines and surface ships should be in separate force structures".

5. The nonstop, non-refueled flight of the *Voyager* aircraft around the globe in December of 1986 illustrates the range enabled by the ubiquity of air.

6. Ref to Karmen limit mentioned previously. Most literature on the subject endorses the above as the limit for air breathing platforms as in practical terms aerodynamic lift above approximately 53–62 km is largely non-existent. Nevertheless, it would be pertinent to note here that maximum operating altitudes of military platforms of the vertical dimension have been progressively evolving from a few thousand feet during World War I, to ten thousand feet during World War II, to hundred thousand feet of the SR-71 and MiG-25s of the Cold War era. As borne out earlier, it was the shooting down of high-altitude aircraft aircraft which provided the initial impetus for going higher and using satellites for safer military reconnaissance.

7. This is best exemplified by the French use of machine guns in the Franco–Prussian war of 1870. The French had developed their machine gun in absolute secrecy and when it was ready, treating it likely an artillery piece, they deployed it in the rear with other artillery, where it could not reach its intended infantry targets, but conversely could be reached and destroyed by the Prussian artillery.

8. The Soviets viewed space as an extension of air, hence their concept of space superiority was embodied within their philosophy about air superiority. For further details on Soviet and Russian concepts driving their military space activities refer to Sqn Ldr KK Nair, "Space Theory and Doctrines", *AIRPOWER Journal,* Vol. 2, No.1, SPRING 2005 (Jan–Mar) pp. 3–6.

9. A typical ICBM weapon system transits the air-space boundary twice during its mission. It spends up to 20% of its transit period in the atmosphere and the rest 80% in space, though during its transit in space it is neither targeting nor attacking targets in space.

10. Refer to Sqn Ldr KK Nair, "Space Theory and Doctrines", *AIRPOWER Journal,* Vol. 2, No. 1, SPRING 2005 (Jan–Mar), pp. 3–6.

11. The extent of misinterpretation can be gauged from the extent that even after four and a half decades of USAF General Thomas White using the above in the context of military operations (USAF DD2-2, Space Operations) and not

conceptual physics, authors have been wantonly casting aspersions and judging the General's beliefs as being "wrong". The logical appropriateness of the General's doctrinal beliefs on the ,use of space and foresight have been repeatedly established in most US conflicts since the Gulf War where air and space assets produced decisive effects. For an outstanding example of such misinterpretation, see Major Samuel L. McNiel, USAF, "Proposed Tenets of Space Power—Six Enduring Truths", *Air and Space Power Journal – Summer 2004.*

12.   The US Air Force Chief of Staff, General Thomas D. White, first used the term air/space on 16 May 1958 in a speech to the Los Angeles Chamber of Commerce. In an article in the August 1958 *Air Force Magazine,* he used 'aerospace' publicly for the first time. This was in reference to Soviet aerospace power. The Air Force used the term more and more frequently in 1958 and adopted "Aerospace Power for Peace" as its slogan in January 1959. For further reading on the term "Aerospace" see Lt. Col. Frank W. Jenning, USAF, "Doctrinal conflict over the word Aerospace", *Airpower Journal, Fall 1990.*

13.   However, it needs to be borne in mind that Eisenhower's policies are a domestic US phenomenon which do not basically bring into question the doctrinal soundness of the concept.

14.   Lt Col Peter Hays and Dr. Karl Mueller, "Going Bodly–Where?" *Aerospace Power Journal,* Spring 2001.

15.   Ibid.

16.   Quoting Major Carl Banner, "Defining Air and Space Power", *Air and Space Power Chronicles,* 11 March 99.

17.   Lt Col Peter Hays and Dr. Karl Mueller, "Going Boldly–Where?", *Aerospace Power Journal,* Spring 2001.

18.   By analogy the Army and Navy also harness the capabilities of air power in support of organic mission; however, only the Air Force exploits holistically the strengths afforded by the environment in support of military objectives.

19.   Apart from numerous academics and military personnel, prominent politicians of the US Senate have also expressed similar views. See Senator Bob Smith (R-N.H.), "The Challenge of Space Power", *Aerospace Power Journal,* Spring 1994.

20.   For details on such like space weaponry, refer to Yorkshire CND briefing "Keep space for peace", June 2004 at http://cndyorks.gn.apc.org/yspace/overview.htm

21.   See General Lance W. Lord, USAF, "Command the future: the transformation of Air Force Space Command" *Air and Space Power Journal,* Summer 2004.

22.   See "Air Force Space System Control Questioned", *Space News,* 8 September

1997, 2. Also see Major Shawn P. Rife, USAF, "On space power separation", *Aerospace Power Journal,* Spring 1999.

23.   See Major Shawn P. Rifle, USAF, "Five myths about the term "Aerospace", *Air and Space Power Chronicles,* Jan. 2001 and John A. Tirpak, "The integration of Air Space", *Air Force Magazine* online, July 2000, Vol. 83, No. 7.

24.   General John P. Jumper, USAF Chief of Staff, to Admiral James O. Ellis, Jr., letter, 23 February 2004, http://www.55srwa.org/0403/04-03011454.html.

Chapter 4

Military Space Theories and Doctrines: A Comparative Overview

National interests and objectives are the fountainhead of a nation's strategies and doctrines. These in turn have a cascading effect on a nation's fundamental, environmental and organisational doctrine thereby providing an insight into the likely courses of action and military force structuring that individual nations are most likely to follow. While national fundamental doctrines would be the fountainhead of most environmental and further organisational doctrines, with regard to space fundamental national ambition would in most cases be tempered by the higher levels of technology, cost and accessibility. The same in turn would have a cascading effect on the likely courses to be followed by most nations. Unlike in the case of the other three environments where environmental doctrines have matured to almost universal applicability levels leading to military force structures being characterised by largely common doctrinal principles tempered by individual capabilities and applications, as regards space no comprehensive universally applicable model exists and most national doctrines and strategies are characterised by a nation's technological, economical and other prowess. Considering this, the deliberations here are limited in scope to the "environmental" level and the manner in which space-faring nations presently view the exercise of military power in/through the medium of space.

Comparative Doctrines

It needs to be borne in mind that as in the case of the other doctrines of land, sea and air, the essential factors of geography, technology and history have a seminal influence on the evolution of space doctrine.

However, the extent of influence of each of these parameters is vastly different. For example, while historical lessons mould the evolution of land doctrines to a far greater extent than geography or technology, maritime doctrines are influenced more by geographical and technological biases, air power doctrines draw from limited historical lessons and as in the case of space are largely a by-product of technology, characterised more by available doctrines being moulded to suit emerging technology rather than vice versa as in the case of land and sea doctrines. This contention is evident in Figure 4.1.

Doctrines and Influences

Doctrine	Influence of History	Influence of Geography	Influence of Technology	Period of Evolution
Land	Highest	Higher	Medium	Dawn of civilisation
Sea	Higher	Highest	Higher	4 centuries+
Air	High	Less	Highest	1 century+
Space	Low	Marginal	Highest	4 decades+

Figure 4.1

It is obvious from the foregoing contention that space's recent evolution and its technological biases distinguish it to a greater extent from land, a lesser extent from maritime and to a significantly much lesser extent from aerial doctrines. Just as air power doctrines draw this basic evolution from maritime doctrines and theories as reflected in the common air and maritime terminology as also missions like sea-control and denial; modern space doctrines build upon air power doctrines premised upon the common characteristics afforded by the vertical dimension like elevation and freedom of movement. The unique characteristic of elevation distinguishes the vertical dimensions of air and space from the horizontal dimension of land and sea with its common envelopment and all-enveloping parvasiveness. Thus modern space doctrines basically build upon basic air power doctrine as reflected in the transmission of the four classical missions of air power into space shown in Figure 4.2.

Additionally, unlike the luxurious time frames moulding the evolution of land and sea doctrines, air power doctrines were evolved in much more compressed time-frames, and with regard to space, the extent of

The Four Classical Missions of Air Power into Space

Role	Typical Air power mission	Contemporary Space mission
Control of environment	Counter air missions	Counter space missions
Applying combat power	Air-based force application	Space-based force application
Multiplying combat power	Airborne combat support	Space-based terrestrial combat support or force enhancement
Sustaining combat force	Support operations	Space support operations

Figure 4.2

compression has only increased. In fact, one of the factors precluding the construct of a comprehensive space doctrine is the pace of technological progress in space capabilities. Humankind's evolution, and the evolution of the environmental doctrines were drawn from contiguous and familiar environs whereas space even now continues to be distant, unknown and not readily accessible The vast expanses of space are yet to be fully comprehended beyond an elemental manner and evolution of operational doctrines for conduct of warfare in space is still immature. The same could also be one of the reasons why most space doctrines are focussed on 'force-enhancement', rather than actual conflict in the realm of space.

Environmental Exclusivity and its Impact

Unlike in the case of air power theories wherein passive uses of air like reconnaissance, targeting, etc. soon gave way to more offensive battles for control of the aerial medium, space, even after four and a half decades is yet to encounter actual conflict characterised by manoeuvring platforms or falling debris as in the case of the other environments. American space doctrines have presumably evolved to the "high ground"; however, these theories are premised upon prevailing theories of air supremacy' or 'air superiority' based upon the premise of dominant strategic force superiority. Space as an environment is less benign to human follies and foibles, it neither provides any "wind beneath the wings" or for that matter gravity to pull down aircraft wreckage or eject aircrew. It needs to be accepted that while war and conflict are inherent characteristics of human nature, outer space and beyond are not forgiving of human trespasses. While planet Earth has tremendous capability for absorbing and tolerating human excesses, outer space and beyond does not, as demonstrated by the effects of experimental explosions on the 'Van Allen' belts and by the cascading effect that a single 'killer ASAT' would entail by destroying an enemy satellite. Fragments of the destroyed satellite would not fall down to Earth as in the case of aerial platforms but would continue orbiting it, hitting other objects like satellites, debris, etc. in turn forming more debris, setting off a chain reaction of destruction that would leave a lethal halo around Earth.

Thus while the US attempts to operationalise elaborate space weaponisation plans and the Chinese attempt to short-circuit the same by developing ASAT weaponry or advertise the development of 'asymmetric strategies' via ASATs, etc., the fact of the matter is that the environment of space is friendly to force enhancement to a larger extent, and force application, counter space operations, etc. to a lesser extent. The classic air strategy chart in case of space would need to be read in the reverse order, terrestrial combat support would obviously be the first priority (history suggests the same) and counter-space campaign would be slotted in the last position in view of the cost, legalities, technical and other issues involved in counter-space operations. This is further validated by the fact that while presently there exist no forms of offensive weaponry in space, the same is primarily because of the technical challenges and cost rather than prevailing treaties or moral and ethical clauses limiting their use. Theories of orbital combat and ASAT weaponry have been in vogue ever since the first *Sputnik* orbited the earth. The Americans focussed on guided missiles, launched from the air and designed to ram the target satellite directly, whereas the Soviets preferred "killer satellites"—orbiting spacecraft armed with shrapnel charges that could disable enemy craft. Both sides also dabbled in nuclear warheads, orbital bombardment systems, etc.; however, all such weaponry was decommissioned and the end of the Cold War killed most of the systems. As recently as 1998, the Rand Corporation had also decisively concluded that while some nations possessed certain capabilities to acquire space weapons, there existed no immediate compelling threat sufficient to drive a country to weaponise space.[1]

Nevertheless a brief examination of the doctrines driving military space programmes in various space-faring nations would definitely be in order considering that cutting across the spectrum of capabilities nations ranging from the US and Russia to China, Israel, etc. are all focused on harnessing space for furtherance of national capabilities of both a military and civilian nature.

Soviet Doctrines

An examination of the Soviet doctrine instead of present Russian doctrines is undertaken in order to understand the unique thought processes that led to the early evolution of Soviet space capabilities which, well after the demise of the Soviet Union, continues to occupy its position as the second most powerful military space apparatus in the world. The edifice of such an apparatus which continues to withstand adversity, slashed budgets, dilapidated infrastructure, etc. and yet possess the capability to resurge is noteworthy for its tenacity.

The political aims of the Soviet state dictated the construct of Soviet military doctrines. Soviet environmental doctrines were a natural offshoot of its overall national military strategy, hence its military space doctrines were also primarily an extension of its overall national military strategy. Soviet space doctrine drew on fundamental Soviet military doctrine and was characterised by an evolutionary rather than revolutionary character like its overall military doctrine which was a derivative of the Stalin era. Thus doctrinal re-evaluations of its military systems complemented rather than superseded existing space doctrines and thereby gave tenacity to its military space programmes, ensuring its survival well beyond the Soviet era and also the near seamless integration into present Russian doctrines. Soviet space doctrines were characterised by a dynamic yet consistent character. The Soviet space programme in turn was characterised by incremental/evolutionary improvements over crude basic structures and reliance upon large quantities of individually less sophisticated systems which provided inherent flexibility, economic viability and an overall impressive capability. While requirements of brevity would preclude an overview of Soviet military doctrine and its impact on Soviet space doctrines, certain overwhelming characteristics of the Soviet military doctrine which decisively fashioned their military space doctrines were as below.

- Soviet military strategy dictates that the best defensive posture is one of military superiority, hence Soviet military doctrine was premised upon using overwhelming force to completely defeat any enemy, at the same time ensuring the protection and survivability of the Soviet Union.
- Soviet military doctrines were also influenced by the fact that the

Soviet Union had suffered the maximum number of casualties (up to 10,000,000) in World War II therefore the deeply ingrained need for a strategy of assured survival as opposed to assured destruction which consequently led to a defensive bias.

- While Soviet military doctrine had a defensive bias, it also recognised the primacy of offensive operations for decisive victory in any military conflict.

- The defensive concepts of Soviet military doctrine were uniformly applicable to all its environmental doctrines of land, sea, air and space as also to the Soviet armed forces and to each of the five services of the Soviet military, viz. the Strategic Rocket Forces (SRF or *RVSN),* Troops of National Defence (*VPVO* or *PVO),* Ground Forces (*SV),* Air Force (*VVS)* and Navy (*VMF).*[2]

- In line with its unique doctrinal requirements, Soviet military space doctrine involved a greater role and involvement for the SRF, PVO followed by the Air Force, the Navy and the Ground Forces.

It needs to be appreciated that contrary to standard Western connotations, the Soviets did not consider offence and defence to be mutually exclusive or even opposing concepts. Ideologically, any weapon developed by them was, by definition, defensive because it was designed to promote "the dictates of history".[3] Thus, while the Soviets consider offence as the most basic component of military operations, a defensive component is an essential part of the overall strategy. Strategic offensive and defensive systems must work synergetically to achieve victory. This is best exemplified by the Soviet interpretation of the term "air defence". To the Soviets, an air defence operation, PVO (*ProtiVovozdushnaya Operatsiya)* is an anti-air operation which has the intention of leading to air supremacy (*gospodstvovozdukhe).* Thus, an inherently defensive operation becomes very offensive in nature and vice versa.

Unlike the Americans, the Soviets at no stage had any misgivings or confusion regarding space being an extension of the vertical dimension of air for the conduct of military affairs. They viewed space as a separate dimension by the strictest definitions of physics, but for operational military purposes they regarded it as an extension of the aerial medium; a fundamental strategic operating medium which, for doctrine and strategic purposes was an extension of the medium of air. They considered space as a part of a Theatre of Military Operations or *Teatr Voyennikh Deystyii* (TVD), hence their objective of space superiority was integral to as well as an extension of the overall objective of air superiority. This had a decisive influence in shaping their space doctrines, space control mission and also space force structuring.

The writings of Marshal V.D. Sokolovsky had a seminal influence on Soviet military space doctrines.[4] The 1968 version of the Soviet Military Strategy outlined the Soviet view on the military uses of space which was based upon the following three paths.

- One, the creation of space weapon systems to assure combat effectiveness for all branches of the armed services.
- Two, preventing other countries from utilising space.
- Three, development of strategic offensive systems to conduct battles in space.

These three paths were broadly analogous with the American doctrines of force enhancement, space control, and force application (aimed at space control unlike in the American context)

The need of a fourth separate mission of 'Space Support', or 'launch-on-demand capability' as in the case of the US was probably not acutely felt or articulated by the Soviets simply because the Soviet strategy of numerous simple satellites and space systems as opposed to a few expensive systems (as in the case of the US) precluded the need to strive for space support capabilities. Soviet design philosophy of satellite standardisation (down to sub-system and component levels) and mass production allowed inherent replenishment, high launch rate capabilites and the ability to absorb greater losses without serious network degradation. In fact, Soviet launch rates have on many occasions been higher than that of the entire world combined and have mostly out-raced its closest competitor—the US—at times to the extent of 5:1 like in 1981 and 1985.

From the foregoing it can safely be inferred that Soviet military space doctrines at an environmental level were also built upon basic air power doctrines as in the case of the US, albeit in its own unique ways. The nuclear backdrop led to a primacy of the 'sanctuary school of thought' with regard to space and the present Russian formation of a separate space force could largely be an example of the continuation of the same. However, Soviet space programmes were not known to be entirely benign and defensive in character. Its testing of the Fractional Orbital Bombardment System (FOBS) and forays into development of ASAT technologies were demonstrative of the popularity the 'high-ground' doctrine enjoyed until the SALT-2 negotiations following which it dispensed with the FOBS systems. By 1971, the Soviets also ended their interceptor tests and focused their efforts from space-control theories to sanctuary. Amongst these three missions, most Soviet efforts were dedicated towards force enhancement and Soviet capabilities spanned the entire spectrum of missions from reconnaissance, communications, to navigation, meteorology, etc. The preponderance of its survivability doctrine was reflected in its overwhelming emphasis on application satellites designed to conduct strategic surveillance. The protection of Soviet tactical and strategic strike capabilities, that is, maintenance of a credible deterrence to enemy aggression was the number one Soviet priority. Fundamental to this objective were the Soviet space-and ground-based sensors which would notify Soviet authorities of an impending attack and allow the exercise of retaliatory forces, implementation of defensive measures, and so on. As a consequence, the Soviets had an elaborate photo reconnaissance programme which was the largest single satellite project in the world and accounted for up to more than a third of its launches. The next most important mission was of possessing survivable communication links, hence communication satellites were the next most numerous satellites. The Soviets had thus invested heavily in mobile satellite communications infrastructure. At the theatre level Soviet satellites provided navigational support for troop deployments, targeting, command, control and communications support via a three-tier space communications network; weather prediction for strike planning; reconnaissance for target location, identification and subsequent strike assessment; and intelligence assessment to provide an order of battle and to guide war managers. Over three-fourth of the Soviet space programmes were assigned to carry out the above objectives either on a dedicated or collateral basis.[6]

The Soviets viewed space as an extension of the vertical dimension hence no distinction was made between the vertical dimensions of air and space. Thus Soviet 'Air Defence' was contiguously extended to include Ballistic Missile Defences as well as space defence. In fact, the PKO's primary mission was to repel any attack emanating from space. Soviet space control objectives were:

(i) Prevention of the use of space by the enemy for military, political or economic gain.

(ii) Unhampered utilisation of space assets to further the Soviet system and goals.

The first objective was targeted towards NATO. The supply lines and communication links from the US to Europe were dependent on satellites, therefore satellites, their links and nodes were key targets for Soviet planners. The second objective concerned the protection of Soviet satellites, their links and nodes, etc. In keeping with these objectives, Soviet space control missions initially laid a lot of emphasis on ASAT and other form of offensive weapon developments in space, but these efforts petered out by the 1970s and the focus was realigned on force enhancement missions. As a natural extension of these thought processes, development of space technology in the Soviet Union and now Russia has proceeded in the following three interrelated directions.[7]

(i) One, the development of space technology to serve wartime needs.

(ii) Two, development of systems for bringing remote sensing information to the lowest level of troop command.

(iii) Three, development of dual-purpose systems that solve the problems of both military and civilian users.

Implementation of these developmental directions was aimed at raising the process of force command and control to qualitatively new levels thereby multiplying combat potential several-fold.

The Chinese Doctrine

The PLA and its component services are not known to use the term "military doctrine" or the classical doctrinal components of roles, missions, tasks, etc. The closest analogue to doctrine is what they term "military science," which in effect links theory and practice. Chinese military science is primarily comprised of:

- Basic military science which includes the fundamental and environmental concepts that govern PLA military operations at various levels of war.

- Applied military theory which deals broadly with application of military force at the various levels of war and is analogous with organisational doctrines.

With regard to space as an environment for the conduct of military affairs, the Chinese have been assiduously studying the effects of space on conventional military capabilities in recent wars and consider space-enabled RMA as an essential component for enhancing their comprehensive national power. The PLA has outlined its mission regarding military space as consisting of two categories.[8] The first is information supporting, and the second battlefield combating which loosely corresponds to missions of 'force-enhancement' and 'counter-space operations' in Western parlance. Its initial aim with relation to military space is within the realm of information supporting. It has further defined this mission as intelligence, navigation, positioning and communication, all of which correspond to missions of force enhancement. This strategy is reflected in the importance it gives to information dominance capabilities or what it terms "informationalisation"[9] followed by development of counter-space capabilities. The same is also reflective in its national policy of taking the road of "composite and leap frog development". Hence China can be expected to attempt exploitation of the entire spectrum of force enhancement missions in a gradual manner and at the same time develop at least rudimentary counter-space capabilities like parasitic ASATs. At a more fundamental level, most literature on the subject in Chinese military journals and Western military writings are suggestive of the following three strategies:

(i) Gradual development of capabilities for exploitation of the entire spectrum of force-enhancement missions afforded by space.

(ii) Integration of the above into terrestrial military capabilities for overall capability enhancement (most Chinese writings on RMA are suggestive of the same).

(iii) Acquisition of ASAT capabilities for a sembalance of 'counter-space' capability so as to possess capability to target US space vulnerabilities thereby providing them with limited deterrence capabilities ("asymmetric advantage") against superior powers.

The foregoing needs to be viewed not in isolation but against China's stated national policy of "Going with the tide of the world's military development and moving along the direction of informationalisation in the process of modernisation, the People's Liberation Army (PLA) shall gradually achieve the transition from mechanisation and semi-mechanisation to informationalisation. Based on China's national conditions and the PLA's own conditions, the PLA persists in taking mechanisation as the foundation to promote informationalisation, and informationalisation as the driving force to bring forward mechanisation. The PLA will promote coordinated development of firepower, mobility and information capability, enhancement of the development of its operational strength with priority given to the Navy, Air Force and Second Artillery Force, and strengthen its comprehensive deterrence and warfighting capabilities".[10] The Chinese fully appreciate the role of space-based systems pivotal to RMA and aspire to adapt themselves to the changes both in the international strategic situation, the national security environment and rise to the challenges presented by the RMA worldwide. China adheres to the military strategy of active defence and is working to speed up its RMA with Chinese characteristics.

The Israeli Doctrine

Israel is a country born in conflict, and its basic security situation has not deviated from that. Survival of the state is an all-pervasive national issue, and the short history of the state has been punctuated by a sequence of serious wars. Israel's survival, therefore depends on its acquisition of 'asymmetric techniques' to offset its inherent disadvantages as also the numerical advantages of its adversaries. As the principal threat has shifted from the proximate land forces of Syria and Egypt to the longer-range missile capabilities of Syria, Iran and Iraq, Israel sees the need to upgrade and modernise her forces across the board.

The need particularly includes extending and upgrading the life of its weapons platform, developing over-the-horizon strike capabilities, space, and ballistic missile defence. In response to the increased militarisation and acquisition of sophisticated military technologies by its adversaries and its own unique needs for survival, Israel has developed a "new war theory" based on which its future missions and force structures of IDF are being developed. The following are its main elements:[11]

- Ensuring that the element of surprise is utilised in any future conflict to achieve a decisive strategic victory for Israel.
- Developing deterrence policies and capabilities, particularly the "deterrence by doubt" factor which aims at intimidating decision

makers in countries that might have tense relations with Israel. This type of deterrence would restrict the movements of Israel's opponents and limit their options and reactions rising from their doubts about the size and type of Tel Aviv's potential response to any military operations.

- Maintaining the technological gap in the fields of armament and military industrialisation between Israel and the Arabs and making this gap as apparent as possible to frustrate the Arabs and convince them that it is impossible to defeat Israel in a future war.

- Possessing strategic weapons systems that can constitute an effective deterrence capable of reaching any target in the Arab region and the Middle East in general.

- Constant readiness to launch a military retaliation whenever necessary that wages total war on an opponent to achieve major objectives and stabilise the situation for long periods of time. A corollary to this readiness, as circumstances require, is to launch a limited retaliatory strike on a small scale that would not jeopardise the peace process.

In view of the foregoing, it may be surmised that the prevailing broad space strategy for Israel would be to initially construct a comprehensive surveillance network of small spy-satellites in LEO to avoid being surprised and at the same time maintain its own 'element of surprise' capability. Its missile defence systems would provide it with its much-needed deterrence capability and the entire military edifice would be supported by its proposed space communications system. Israel's strategy of using small satellites coupled with its proficiency at orbital detuning as also micro and nano-sat technology would provide it the capability to fulfil its force-enhancement and space-support missions, in addition to an incidental 'ASAT' capability, if the need arose. Israel fully comprehends the capabilities and opportunities offered by space for enhancing its ability to achieve its security objectives. Israel's strategic doctrine dictates that space-driven RMA is critical to its future military operations.

Israeli planners view space as a natural extension of air power to be harnessed synergistically to its advantage as the balance of power in the region shifts increasingly to ballistic missiles. Regular Israeli pronouncements on the subject like "the ability to utilise space is the cornerstone that supports air power", "Whoever cannot achieve this…will not be in the forefront of modernisation", etc. by former Israeli Air Force commander Major General Eitan Ben-Eliahu, an ardent space advocate as also pronouncements by its present commander Major General Dani Haloutz that[12] "In summary, air and space power complement each other operationally through command and control of air missions, targeting, intercepting ballistic missiles, providing weather information over long-distance areas, and intelligence mission escorting. Building up of military forces should be based on exploring and understanding the trade-offs between air and space requirements—each can be operationally integrated to provide greater capabilities than perhaps might be available separately", serve to validate the view that Israel fully comprehends the full import of air, space and aerospace power. Once its finances stabilise, it would make all-out efforts to boost its comprehensive military power with a generous element of space-based capabilities.

American Space-power Theories and Doctrines

Worldwide, and in particular American writing on the subject has been profuse ever since the demonstrated effects of space in the Gulf War. Space power theory is also growing. Compared to the works of authors such as Clausewitz, Mahan, and Douhet, space power theory is relatively embryonic. The earliest published works date back to just about four decades. However, most of the meaningful writing on space power occurred in the last decade (not surprisingly, after the Gulf War). Nevertheless, for the sake of brevity, some of the popular American theories are outlined in the following paragraphs.

Lupton's Theory[13]

Most literature on the subject endorses the fact that David Lupton's four-space military doctrinal approaches provide an important and comprehensive way to analyse the strategic rationale driving military space activities, in addition to providing a useful analytical framework for military applications of space. Each of the four approaches, namely sanctuary, survivability, control and high ground suggests a focus, employment strategy, wartime mission and preferred organisation for space forces.

The sanctuary view of space doctrine believes that the realm of space should not be weaponised. It emphasises that space systems are ideal for monitoring military activity, providing early warning to reduce the likelihood of surprise attack and serving as a means to enable and enforce strategic arms control. The basic tenet of the sanctuary doctrine is that space surveillance makes nuclear war less likely. It is closely linked to the deterrence theory and the assumption that no meaningful defence against a nuclear attack by ballistic missiles is possible.

The survivability doctrine emphasises broad utility for military space systems, not only at the strategic level emphasised in the sanctuary doctrine, but also at the tactical level of space support to the war fighter that has emerged as the most important force enhancement mission since the end of the Cold War. It differs from the sanctuary doctrine because it highlights space system vulnerabilities and questions whether space can be maintained as a sanctuary due to ongoing technological improvements in systems such as ASAT weapons. The origins of this approach can be traced back to the late 1970s and early 1980s when Soviet testing of ASAT weaponry led to beliefs that space forces are less survivable than terrestrial forces. The survivability approach consequently argued that space forces must not be depended upon for providing various functions such as communications, navigations, etc. because they may not survive in actual conflict. This gave rise to the space control doctrine.

The space control doctrine is analogous with concepts of air superiority and considers space as another military theatre with the military objective of seeking control over the space environment. Thus, the control doctrine sees space as similar to other military environments and argues that both commercial activities and military requirements dictate the need for space surveillance, as well as offensive and defensive counterspace capabilities.

The high ground doctrine had its origins in President Reagan's SDI which advocates space-based ballistic missile defence. It argues that space is the dominant theatre of military operations and is capable of affecting terrestrial conflict is decisive ways. As a primary example of such capability, the high ground doctrine points to the potential of space-based BMD to overturn the dominance of offensive strategic nuclear forces. This doctrine also believes that space has the ability to be the critical factor in determining the outcome of battle. It uses the analogy that domination of the high ground ensures domination of the lower areas, hence in future space forces will dominate terrestrial forces. The US has over the years presently evolved to this school of thought.

Mantz' Theory[14]

The foregoing theory treats space at the level of an environment only for prosecution of actual combat roles and missions and is more of a space combat theory which incompletely addresses the other missions possible through space. In fact, it almost entirely omits the most important role of force enhancement. It looks more at the application of fire power in space and omits even important aspects of information dominance, etc. enabled by space. Mantz' theory looked at fire power in space. His publication predicted space power to progress in much the same manner that air power has—to the point of being a decisive force by applying long-range, strategic fire power (including bombardment) through the sky. Mantz' discussion of space strike operations gives amplification to the idea of decisive space power as a potential war winner. It also gives specific details to accomplish space control in a weaponised space environment. Some of the important aspects of the above theory are as follows:

- Space strike systems can be employed decisively by striking earth forces, both independently and jointly.
- Space strike systems can be employed decisively in war when the enemy's essential means for waging war (industry, transportation, and communications) are vulnerable to attack from space.
- Space strike systems can be employed decisively by striking at the decision-making structure (leadership and command and control) of the enemy.
- Space strike systems can deter hostile actions by holding forces, decision making (leadership and command and control), infrastructure (industry, transportation, and communications) at risk.
- Space denial systems can be employed decisively by denying enemy access to space-derived data.
- Space denial systems can be employed decisively by physically denying enemy access to space.
- Space protection systems can be employed to assure friendly access and use of space.
- Total space control (the combination of space denial, space

protection, and passive space defence measures) is neither achievable nor necessary.

● Space combat power must be centrally and independent controlled.

● Space power is not intrinsically linked to air power.

The types of space combat missions envisaged are shown in Figure 4.3.

Types of Space Combat Missions

Space Denial Ops	Space Strike Ops	Space Protection Ops
Earth-to-Earth attacks	Space-to-land/subterranean attacks	Protecting against Earth-to Earth attacks
Earth-to-space attacks	Space-to-sea/undersea attacks	Protecting against Earth-to-space attacks
Space-to-space attacks	Space-to-air attacks	Protecting against space-to-space attacks
Space-to Earth attacks		Protecting against space-to-Earth attacks

Figure 4.3

Gravity Well Theory[15]

The aforementioned was a multi-author effort and was based upon exploiting the combat advantages afforded by gravitational energy as one went higher. It was a virtual extension of the 'high ground' theory of land warfare into space. The high ground theory builds upon the premise that by commanding the hill one could control the surrounding country and influence battle in one's favour, thereby winning the war. Harry Stine captured the military significance of the gravity well theory in two axioms:

(i) Control of the moon means control of Earth, and

(ii) Control of the L-4 and L-5 libration points means control of the entire Earth–Moon system (see Figure 4.5).

From here, Stine stated that a military commander has the ability to permit or deny passage of space traffic, to deny the use of other military or commercial orbital areas to others, to launch strikes against *any* target on Earth, on the moon, or in Earth–Moon space, or detect and take action against any threat originating anywhere in the Earth–Moon system.

According to Stine, shots fired down the gravity well will travel faster than shots fired up the well. Thus, the higher person has an energy advantage (he will not have to fire his shot as fast in order to maintain the same speed as one shooting up the well). The higher person also has a manoeuvring advantage (all things being equal, the higher person will have more time to observe the attack and dodge the shot than his opponent). Therefore, to maintain a military advantage in space, one must remain higher up the gravity well than his adversary.

Stine goes on to discuss exactly where the "highest" ground in space is. According to him, a libration point (also called Lagrangian point after Joseph Louis Lagrange, the French astronomer who suggested the exercise of these points around 1800) is a location where gravitational forces are theoretically in perfect balance. Three of the five libration points are considered unstable because the Moon's non-circular orbit and the Sun's gravitational pull adversely affect these points. The last two libration points (L-4 and L-5), also known as Trojan libration points, are not affected by these phenomena and are thus considered stable. At these two locations a body in space would theoretically require little to no energy to sustain its position in the Earth–Moon system. Therefore, L-4 and L-5 are the highest ground in the Earth–Moon system.

Gravity Wells

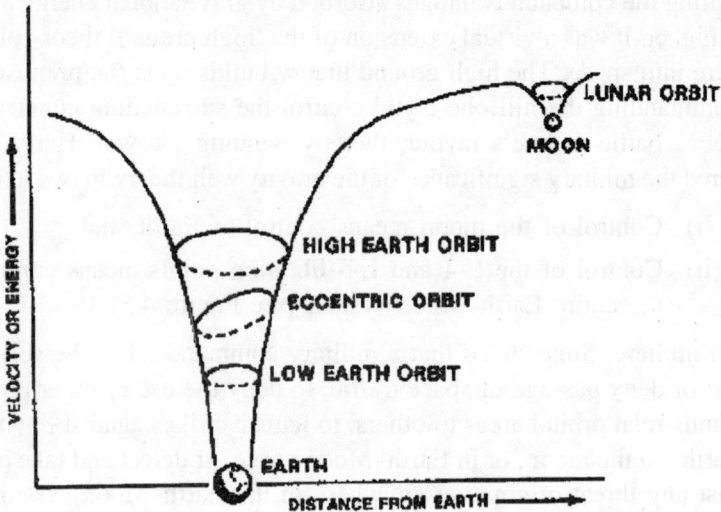

Figure 4.4

Apart from Lupton's theory, the remaining theories have not found much favour and have mostly been a subject of academic discussions and debate rather than actual application.

US Space Doctrine

American space doctrines have always been characterised by a preponderance of Air Force doctrines in exploiting space for military purposes. This has been the trend since 1957, when, following the Soviet *Sputnik* shock, the Americans entrusted the responsibility for military operations in space to its Air Force and by 29 November 1957, the first Air Force space doctrine was announced by the Air Force Chief of Staff, General Thomas D. White. This included the ideas that space power would prove as dominant in combat as air power; that there is only one operational medium of aerospace since there is no distinction between air and space; and the Air Force should have operational control over all forces within this medium. General Thomas D. White first used the word 'aerospace' in 1958, and the concept that air and space form a seamless operational medium has been the foundational component of American

Lagrangian Points in the Earth Moon System

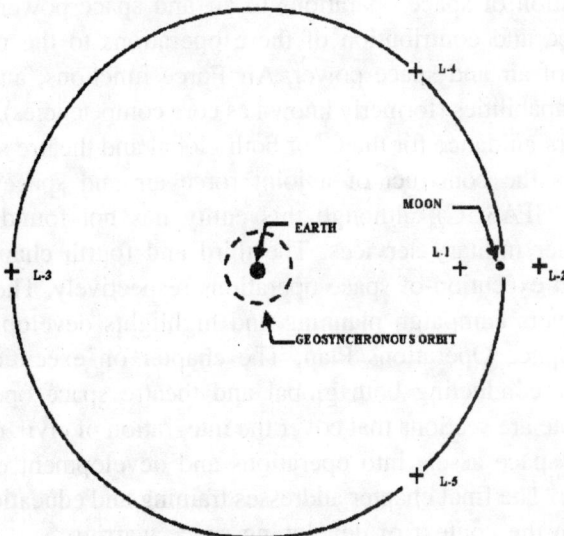

Figure 4.5

thinking about space ever since. This is reflected in the US Air Force's adjustment of its doctrine to accommodate space-based capabilities so as to enable optimal exploitation of the capabilities afforded by space. Air Force stewardship of space in the US is apparent considering that it provides 90 per cent of the military's space budget, as also 93 per cent of space personnel. Despite a post-Cold War drawdown that has seen all of the services reduce their size by roughly a third, the Air Force satellite force has increased by 25 per cent since 1991.[16]

Currently US space operations doctrines have been codified into operational-level space-operations doctrine in the Air Force Doctrine Document (AFDD) 2-2, *Space Operations,* and Joint Publication (JP) 3-14, *Joint Doctrine for Space Operations,* respectively. AFDD 2-2 and JP 3-14 are similar both in scope and content. The Army also has a rather dated document, Field Manual 100-18, *Space Support to Army Operations,* 20 July 1995 which has not been revised since and hence is not deliberated upon.

Current Air Force operational-level space doctrine resides in AFDD 2-2, last revised in 2001. This document provides significant detail in the areas of command and control (C^2) of space forces as well as the planning and execution of space operations. The first chapter serves as a primer on the contribution of space operations to air and space power, examining the relevance and contribution of these operations to the principles of war, tenets of air and space power, Air Force functions, and Air Force distinctive capabilities (formerly known as core competencies). The second chapter offers guidance for the C^2 of both global and theatre space forces. It introduces the construct of a joint force air and space component commander (JFASCC), although this entity has not found acceptance with the other military services. The third and fourth chapters discuss planning and execution of space operations respectively. The chapter on planning covers campaign planning and highlights development of the Air Force Space Operations Plan. The chapter on execution provides guidance for conducting both global and theatre space operations. Of particular note are sections that cover the integration of civil, commercial, and foreign space assets into operations and development of the space tasking order. The final chapter addresses training and education for space operations in the context of developing space warriors.

JP 3-14, which treats joint space-operations doctrine, was published in 2002 after undergoing development for well over 10 years. Though a fairly recent document, it needed revision as soon as it appeared due to the merger of US Strategic Command (USSTRATCOM) and US Space Command (USSPACECOM). Divided into five chapters and eight appendices, JP 3-14 includes material similar to that of its service counterparts. The first chapter provides an overview of military space operations and the operational considerations for space. The second, which covers space organisations and their responsibilities, requires significant revision because of the creation of the new USSTRATCOM. The third chapter offers guidance for the C^2 of space forces, focussing primarily on global space forces but including limited guidance on command and support relationship for theatre space operations. The fourth discusses military space operations in the context of the principles of war and the four mission areas for space (control, force enhancement, support, and force application). The final chapter discusses deliberate and crisis-action planning for space operations. The appendices provide a tutorial on several topics, including intelligence, surveillance, and reconnaissance (ISR); integrated tactical warning and attack assessment; environmental monitoring; communications; position, velocity, time, and navigation; and orbital characterisation.[17]

In spite of the above, there exists no doctrinal agreement on a comprehensive all-enveloping doctrine capable of addressing all the aspects of space. Vigorous discussions and deliberations on the subject are rife with no commonly accepted theory or doctrine in sight. However, by all available accounts and literature on the subject, the Americans are presently moving in to the "high ground" school of thought to consolidate their gains in space.

Conclusion

In view of the foregoing, it is apparent that in the absence of any compelling countervailing theory, most nations have built their space doctrines upon the basic air power doctrine. Basic air power missions form the framework for building upon space power capabilities and in spite of semantic disagreements on the terminology, most nations in their unique fashion build upon the same depending upon their individual requirements, capabilities and threat perceptions. Thus while the Americans are gradually evolving to space-control doctrines, the Russians are slipping back into the broad equivalent of the sanctuary doctrine and limiting their programmes to force-enhancement missions, the Chinese on the other hand are attempting a mixed strategy aimed at exploiting the entire spectrum afforded by force-enhancement missions and also a leapfrog into counter-space capabilities. The limiting riders in most cases, i.e., cost and technology, however, continue to be the same, explaining the wide diversity of force structuring being undertaken within broadly common doctrinal applications.

Notes and References

1. Bob Preston et al., *Space Weapons Earth Wars* (Rand MR-1209-AF), Ch. 6, p. 99.

2. RVSN stands for *Raketnye Voyska Strategicheskogo Naznacheniya,* VPVO for *Voyska Protivovozdushnoy Oborony,* SV for *Suchoputnyye Voyska,* VVS for *Voyenno Vozdushnyye Sily,* and VMF for *Voyenno Morsokoy Flot.*

3. Lawrence E. Stellmon, "A Comparison of US and Soviet Programs", *Strategic Studies Report abstract,* National Defence University, February 1987, Ch. 2, p. 2.

4. See VD Sokolovskiy, *Soviet Military Strategy,* 3rd edn, trans, Harriet Fast Scott (NY: Crane, Russak 1975), pp. 84–5.

5. Nicholas L. Johnson, *Soviet Military Strategy in Space,* London: Jane's 1987, p. 198.

6. Ibid., p. 199.

7. Valerity A. Menshikov, *Military Uses of Space,* Encyclopedia of Space Science and Technology, Vol. 2, A. John Wiley & Sons, Inc. Publications, 2003, p. 121.

8. You Ji, *The Armed Forces of China* (NY: I.B. Taurus, 1999), p. 84.

9. For a more elaborate explanation on "informationalisation", refer Chinese government White Paper, "China's National Defence in 2004", Ch. 3.

10. Ibid.

11. Jameel Al-Din Husayn, "Israel Peace and Arms", *Cairo Rose Al Yusuf,* 21–27 August 1999, FBIS document ID: FTS 19990826001012.

12. Maj. Gen. Dani Haloutz, Commander of the Israeli Air Force quoted in *Air and Space Strategy for Small Powers,* Rand CF-177-FIAS, Ch. 6, p. 156.

13. David E. Lupton, *On Space Warfare: A Space Power Doctrine,* Alabama, Maxwell Air Force Base: Air University Press, 1983.

14. Michael R. Mantz, *The New Sword, A Theory of Space Combat Power,* Maxwell AFB, AL: Air University Press, May 1995.

15. G. Harry Stine, *Confrontation in Space,* Englewood Cliffs: NJ: Prentice-Hall, Inc. 1981, p. 58.

16. James Kitfield, "Space and Air Force", *Air Force Magazine,* Vol. 81, No. 2, February 1998.

17. Available at www.dtic.mil/doctrine/jel/service_pubs/afdd2_2_1.

Chapter 5

Examining Space Law and its Military Implications

"The law is an ass—an idiot…and the worst I wish the law is that his eye may be opened by experience—by experience."

— Charles Dickens, *Oliver Twist,* Chapter 51, p. 489.

The Genesis

As in the case of air law, legislative endeavours on space preceded the actual technology involved and while the *Sputnik* was launched on 04 October 1957, the origins of space law can be traced back to as early as 1934 or even before,[1] when Eygene Korovin, professor of international law at the Institute for Soviet Law in Moscow, published his article on the international legal aspects of the stratosphere, "La conquête de la stratosphere et le droit international" in *Revue Gènèrale de Droit International Public* (vol. 41, pp. 675–686). Korovin was widely regarded as the father of scientific international space law in the Soviet Union. With surprising clairvoyance he foresaw that aviation in outer space ('stratosphere') would develop along the same lines as in the lower air layers. The steps defined in 1934 by Korovin appear to be prophetically sound: after the initial stage of adventure came the military involvement during the Cold War and today satellites in outer space constitute an all-pervasive force for promotion of military, economic, social and scientific development.

Such an abundance of resources and developmental potential inherent in space-based activities also carries immense potential for individualistic appropriation by state and non-state actors, and consequently immense potential for competition, rivalries, disagreements and conflicts. It was in recognition of this predicament that world bodies like the UN put forth certain legalities, principles and norms with respect to outer space.

Historical legal analogies usually form the basis of fresh law formulation, hence to fill the vacuum of space law most of the initial efforts at creating laws in outer space were premised upon three analogies—air, high seas, and Antarctica. Each of these analogies suggested a distinct approach to the regulation of space. The air and high seas analogies implied treating outer space as open to forms of military activity accepted under general international law, while the Antarctic analogy suggested treating outer space as "off limits" for all military activity. These three analogies are briefly dwelt upon so as to appreciate and examine its conceptual applicability and contextual currency in the past, present and future.

(a) The Air Analogy

The air analogy supported notions of state control over all activity above its territory. It implied that the same rules regarding military activity that prevailed within a state's own domain, including its air space, should be applied to outer space. These rules included the right to construct and maintain weapons and armed forces, and to use armed force against unauthorised intruders in self-defence. For example, military aircraft intruding upon national airspace can be shot down whereas civilian aircraft can be escorted or forced down, as the case may be.

(b) The Sea Analogy

The basic rationale for this principle was set forth by the Dutch Jurist, Hugo Grotius (who eventually became known as "the father of international law") in 1609 in his famous *Mare Liberum*. This analogy supported the notion of freedom of the seas. Freedom of the seas is the principle that outside its territorial waters, a state may not claim sovereignty over the seas except with respect to its own vessels. He came up with a concept which, translated, meant "Free Seas". It meant that the sea should be open to all nations. The seas cannot be appropriated by one sovereign, or even by a number of them. The seas and oceans are the province of all mankind. It was from this that concepts related to the "common heritage of mankind" were derived.

(c) The Antarctic Analogy

This analogy, available after completion of the Antarctic Treaty, suggested non militarisation of an entire area. The treaty stated that Antarctica shall be used "for peaceful purposes only", and defined this to mean a prohibition on all military activities, inclusive of

- The establishment of military bases and fortifications
- Carrying out military manoeuvres
- Testing any type of weapons.

This entailed far more comprehensive limitations than those which prevailed within the state domain or on the high seas. It banned even forms of military activity regarded as defensive under the UN Charter. Nevertheless, it allowed for military personnel and equipment to be used for scientific research and for peaceful purposes.

The fundamentals of space law share corollaries both with maritime law and aeronautical law, some prominent examples include determining nationality and registration, assessment of liability, and operational and personnel safety. However, not all international rules governing the air and sea are directly applicable to space. The unique aspects of the environment of space preclude the use of any particular set of analogous legal proscriptions to accommodate the multitude of potential problems there. Nevertheless, the analogies are often similar enough to allow formulation of acceptable rules for the space environment. For example, akin to the concept of freedom of the high seas which is considered a tenet of maritime law, the Soviet Union's launch of *Sputnik* established the precedent of free passage through space for orbiting satellites. Since there was no objection to over-flight by *Sputnik* or subsequent satellites, it became a legal custom whereby nations accepted the proposition that they did not retain sovereignty of the outer space above their territories.

Prior to even the launch of the *Sputnik,* the world community inclusive of the then prevailing space powers of the US and USSR overwhelmingly favoured the use of space for peaceful purposes. Further moves to ensure that 'outer space be used exclusively for peaceful and scientific purposes and for the benefit of mankind' included the joint submission by four Western powers (Canada, France, the United Kingdom and the United States) to the United Nations Disarmament Commission, calling for a study on an inspection system that would assure that objects launched into outer space would be used exclusively for peaceful and scientific purposes. Adopted by the General Assembly, this became the first United

Nations resolution on outer space, and the first time the phrase 'exclusively for peaceful purposes' would be used in the authoritative United Nations text.[1]

The thirteenth session of the General Assembly, held in 1958, provided a forum for the debate on 'Questions of the Peaceful Use of Outer Space'. During this session the term 'peaceful' was used as an antonym to 'military'. Sweden appealed to fellow Member States to 'safeguard outer space against any military use whatsoever' and the Soviet Union put forward a proposal to ban the use of outer space for military purposes. The General Assembly adopted resolution 1348 (XIII), which recognised the 'common aim' of humankind that outer space 'should be used for peaceful purposes only'.[2]

Starting in 1958, the General Assembly passed a number of resolutions establishing basic concepts for a space law regime. These concepts include: that international law, including the UN Charter, is applicable to outer space and celestial bodies; that outer space and celestial bodies are free for exploration and free from national appropriation; that principles such as state and corporate responsibility, ownership, and control be applied to the operation of space vehicles; and that arms control principles are applicable to space. In 1967 many of these concepts were incorporated in what is the basic document of current space law: The Outer Space Treaty (OST)-1967, supporting treaties in 1968, 1973 and 1976 established principles of astronaut and space object control, liability and compensation for damage caused by space objects, and the registration of all space objects with the UN.

Space Law

Space law in its present avatar is an amalgamation of many fields of law such as international law, the law of torts, contracts and property, etc. and is primarily a derivative of various treaties and conventions, customary international law and recognised general principles. It deals with the rights of many different and often competing or incongruous entities such as government agencies, private companies, and individuals from each of the world's nations. As described by the UN, space law is "the body of law applicable to and governing space-related activities". The term "space law" is most often associated with the rules, principles and standards of international law appearing in the five international treaties and five sets of principles governing outer space which have been elaborated under the auspices of the UN. However, space law also includes international agreements, treaties, conventions, rules and regulations of international organisations (e.g. the International Telecommunications Union), national laws, rules and regulations, executive and administrative orders, and judicial decisions.[3] The primary goals of space law (as elaborated upon by the UN) are to ensure a rational, responsible approach to the exploration and use of outer space for the benefit and in the interests of all humankind.[4] To this end, space law attempts to address a variety of diverse matters, such as military activities in outer space, preservation of the space and Earth environment, liability for damages caused by space objects, settlement of disputes, protection of national interests, rescue of astronauts, sharing of information about potential dangers in outer space, use of space-related technologies, and international cooperation.

Thus, space law could be broadly defined as "the body of law applicable to state, non-state and private individual entities governing the conduct of activities in outer space". Generally speaking, the prevailing laws of outer space can be divided into two areas:

- **International space law,** which governs the activities of States and international intergovernmental organisations and
- **Domestic space law,** which governs the activities of individual countries and their nationals.

International space law provides a commonly accepted framework of rules and guidelines for nations to use in dealing with one another. Recognised sources of international space law include treaties and conventions in force, international custom based on accepted practices, and general principles of law. Unlike domestic law, which has clearly defined enforcement mechanisms and specific penalties for non-compliance, observance and enforcement, international law is often determined by the willingness of nations to cooperate. Hence, it also needs to be borne in mind that in actual practice the following four basic premises generally guide its application:

- Arms control treaties bind only those who agree to them.
- Activities not expressly prohibited by treaty or agreement are assumed to be permitted.
- Treaty provisions are often subject to interpretation according to the needs of the states that are constrained by them.

- States may withdraw from a treaty in accordance with the treaty's provisions for doing so or as necessary to defend itself during hostilities, unless the treaty specifically requires otherwise.

International Space Laws

The fundamental principles of public international space law can be found primarily in the following five multilateral treaties and principles drawn under the auspices of the UN in addition to other agreements and resolutions, shown in Figure 5.1.

Fundamental Principles of Public International Space Law

Treaties	Principles
• OST, 1967	• Governing activities of states in exploration and use of outer space, 1963
• Rescue Agreement, 1968	• Governing use by states of artificial earth satellites for indirect TV broadcast, 1982
• Liability Convention, 1972	• Relating to remote sensing, 1986
• Registration Convention, 1975	• Relevant to use of nuclear power sources, 1992
• Moon Treaty, 1979	• International cooperation in exploration and use of space

Agreements

- 1963 NBT–Treaty Banning Nuclear Weapon Tests in the Atmosphere, in Outer Space and under Water.
- 1974 BRS–Convention Relating to the Distribution of Programme-Carrying Signals Transmitted by Satellite.
- 1971 ITSO–Agreement Relating to the International Telecommunications Satellite Organisation (ITSO).
- 1971 INTR–Agreement on the Eastablishment of the INTERSPUTNIK International System and Organisation of Space Communications.
- 1975 ESA–Convention for the Establishment of a European Space Agency (ESA).
- 1976 ARB–Agreement of the Arab Corporation for Space Communications (ARABSAT).
- 1976 INTC–Agreement on Cooperation in the Exploration and Use of Outer Space for Peaceful Purposes (INTERCOSMOS).
- 1976 IMSO–Convention on the International Mobile Satellite Organisation.
- 1982 EUTL–Convention Establishing the European Telecommunications Satellite Organisation (EUTELSAT).
- 1983 EUM–Convention for the Establishment of a European Organisation for the Exploitation of Meteorological Satellites (EUMETSAT).
- 1992 ITU–International Telecommunication Constitution and Convention.

Figure 5.1

The international legal regime also includes various agreements covering the commercial uses of space, such as rights to use the geostationary orbit and agreements establishing intergovernmental organisations (for example, the Intergovernmental Agreement on the International Space Station, the International Telecommunications Union, and the World Meteorological Organisation). Up to 72 resolutions have been passed on the subject since 1958.[5] It is pertinent to note that while the provisions of the aforementioned treaties under international law are binding upon states which have ratified them, the principles enumerated above are regarded by modern international law systems as legally non-binding although they have the legal status of General Assembly resolutions. They (the principles) serve only to provide generally accepted rules and standards by which States may, and very often do, govern their space-related activities. Nevertheless, some of these principles address quite sensitive issues, e.g. remote sensing of Earth or the use of nuclear power sources in space. Similarly, UN resolutions are also not binding and have no direct power to regulate space activities.

Domestic Space Law

The foregoing are mostly national endeavours developed under the broad guidelines of international space law aimed at organising national space efforts as well as regulating and protecting national space efforts, assets and interests. Domestic space laws are generally designed to protect and promote national interests within the overall framework of international space law; however, some of the overwhelming reasons for enacting such laws are as under:

- As per international space law, states (nations) bear international responsibility for national activities in outer space, including the moon and other celestial bodies, whether such activities are carried out by governmental agencies or non-governmental entities, and for ensuring that national activities are carried out in conformity with the provisions set forth in the treaty. The Treaty further states that the activities of non-governmental entities in outer space, including the moon and other celestial bodies shall require authorisation and continuing supervision by the appropriate State Party.[6] Therefore, domestic legislation and licensing restrictions are one way in which States can accept the above obligations and apportion the risks of such activities.

- States have an interest in assuring the efficient use of space without harmful interference.[7] Licensing restrictions can help meet this goal, as can management of radio frequencies and the GeoStationary Orbit (GSO) through domestic implementation of the international regime under the International Telecommunications Union (ITU). States also have an interest in ensuring that the use of space does not threaten their national security. Regulation and control of domestic space activities is generally done by licensing regimes which are a powerful tool to address this concern.

The fact that national security is a major factor in the decision to grant each of the above types of licences is obvious when one considers the purposefully broad applicability of the laws. In addition, most States openly include national security or national interest as a factor in deciding whether or not to grant a license to engage in space activities. For example:

- United States Commercial Space Act 1998 can prevent a launch if it "would jeopardise the public health and safety, or any national security interest or foreign policy interest" of the US.[8]

- Russian Federation's national space law can shut down operations of readying for launch, or other operations at the site of conducting space operations if: carrying them out threatens the health and safety of people, and also the state interests and security of the Russian Federation; operations are being conducted without a licence, or in violation of the conditions of its use.[9]

- UK's national space law contains provisions for not granting a licence for impairment of its national security, inconsistency with UK's international obligations, concerns of public health, safety of persons or property.[10]

- Australia's Space Activities Act of 1998 can refuse a licence "for reasons relevant to Australia's national security, foreign policy, or international obligations". The Act applies to domestic launches and overseas launches by domestic entities.[11]

- South Africa's Space Affairs Act takes into account the minimum safety standards, the national interest of South Africa, as well as international obligations and responsibilities.[12]

Presently, States which have national law and legislation governing space-related activities inter alia include Argentina, Australia, Canada, Finland, France, Germany, Hungary, Indonesia, Japan, New Zealand, the Philippines, Republic of Korea, Russian Federation, Slovakia, Sweden, South Africa, Tunisia, Ukraine, the UK, and the USA. It is indeed surprising that while countries with nil or nascent space capabilities like Tunisia, Finland, Slovakia, etc. have elaborate domestic space legislations in place, Asian space-faring nations like China, India or Israel have no such domestic space legislation in place. Whether this is by default or design is a moot point beyond the scope of prevailing deliberations.

Space Law in the Indian Context

As regards international space law, India has always been an active participant right from the formative years of international space law. It has ratified four out of the five UN treaties, is a signatory to the "Moon treaty" and party to all the five UN principles enumerated above. India's position with regard to the other space-faring nations[13] with respect to fulfilling international obligations is given in Figure 5.2.

India's Position Versus Other Space-faring Nations

Country	International Space Treaties					Significant International Agreements				
	OST	ARRA	LIAB	REG	MOON	NTB	BRS	ITSO	IMSO	ITU
USA	R[14]	R	R	R	–	R	R	R	R	R
RUS	R	R	R	R	–	R	R	R	R	R
CHN	R	R	R	R	–	R	–	R	R	R
JPN	R	R	R	R	–	R	–	R	R	R
IND	R	R	R	R	S	R	–	R	R	R
ISRL	R	R	R	–	–	R	S	R	R	R
ESA	–	D	D	D	–	–	–	–	–	–

Figure 5.2

From Figure 5.2 India's commitment to international peace and treaties promoting peaceful uses of space is apparent, considering the fact that amongst space-faring nations, India, in spite of a planned moon mission, is the only signatory to the immensely undersubscribed "Moon treaty". Indian compliance with all the space treaties, principles and relevant agreements is truly outstanding. India has also been an active participant in formulating legal principles relating to the current issues of international space law, such as use of Nuclear Power Sources in Outer Space; definition and delimitation of outer space; character and utilisation of geostationary orbit; the concept of "Launching State"; etc.

However, India's domestic space law endeavours are woefully inadequate or to be less poetic and more accurate, absolutely absent. India has no form of domestic legislation in place for regulating national space activity, or to discharge its international or domestic obligations with regard to liabilities and other issues, or even for guarding against jeopardising the security of its citizens and other national assets or even, for that matter ensuring that the use of space does not threaten its own national security. India would face problems of international responsibility in the event of not discharging/violating its obligations under the international conventions and may face the problem of liability for acts not prohibited by international law. Presently no licensing regime exists within the country, which is surprising considering that India is in the thick of privatisation and globalisation. ISRO itself is willing to privatise some of its activities, while the Antrix Corporation is engaged in commerce. This entails addressing the legal issues of liability and compensation for injury/harm caused by space objects, registration and insurance of space objects, procedures and forums for settlement of claims, etc. Nevertheless, some analysts on the subject contend that the above lacunae are covered to a certain extent since the Indian State and its instrumentalities are subject to Constitutional mandate given in Article 51(c) of the Indian Constitution that they should foster respect for international law and treaty obligations.[15] The fact of the matter is that India is in desperate need of domestic space legislation to fulfill its domestic and international obligations with respect to space law.

Space Legislations Regulating Military Activities

An assessment of the law regarding the military use of outer space must begin with reference to the United Nations Charter, binding upon every country in the world. Although adopted well before *Sputnik,* the Charter knows no geographic limitations: it is fully applicable to the behaviour of states on, under, and above the planet. Articles 2 and 51 of the UN Charter are the most significant in terms of military operations in space. Article 2(4) of the UN Charter prohibits the threat or use of force in international relations, while Article 51 codifies the right of self-defence in cases of aggression involving the illegal use of force by another state(s). Further, the Charter contains a unique supremacy clause: in the event of a conflict between the Charter and any other treaty (whether pre-existing or subsequently concluded) the obligations of the Charter shall prevail. Apart from the previously mentioned international treaties, principles and agreements, there is a number of space-specific treaties regulating or constraining military activities in space which primarily include the Limited Test Ban Treaty (LTBT)—1963, Outer Space Treaty (OST)—1967, Moon Treaty—1979. Certain bilateral treaties of US–Soviet vintage like the ABM, SALT, and START had a seminal influence on limiting military space activities, but are no longer relevant in the present unipolar context,[16] or have a drastically reduced influence on the subject. Treaties addressing technical means of verification and missile defence like the Strategic Arms Limitation Talks (SALT) treaty of 1972 have been taken over into the Intermediate Range Nuclear Forces (INF) Treaty, which is of indefinite duration, and into the Treaty on the Reduction and Limitation of Strategic Offensive Arms (START-1), which has been extended to 2009. The intent of this non-interference measure is to preserve from attack or interference technical means of verifying treaty compliance, including space-orbiting means. It would be a violation of the provisions on non-interference with national means of verification in the INF and START-1 treaties to use weapons against any early warning, imaging, or intelligence satellite and, by extension, against any ocean surveillance, signals, intelligence, or communications satellite of the United States or Russia. This obligation was made multilateral in the Conventional Forces in Europe (CFE) Treaty, which has 30 NATO and East European participants and is of unlimited duration.

An additional set of limitations was imposed by the 1977 convention on the Prohibition of Military or Any Other Hostile Use of Environmental Modification Techniques (ENMOD) which bans the use of certain environmental modification techniques aimed at changing the dynamics, composition, or structure of outer space, and the International Telecommunications Convention, which contains provisions relating to space communications.

In an attempt to control potential dual civil–military use of space launchers, stringent non-proliferation control regimes like the Missile Technology Control Regime (MTCR) and the Wassenar arrangement have also been put into place. Other military peace-time treaties concerned with regulating weapons with potential space applications include, for example, the 1963 Partial Test Ban Treaty, the 1972 Biological and Toxins Convention, the 1970 Nuclear Non-Proliferation Treaty and the 1992 Chemical Weapons Convention. Finally, the treaties which have an impact on space security during times of armed conflict include the corpus of international humanitarian law composed primarily of The Hague and Geneva Conventions, also known as the law of armed conflict. These treaties regulate the means and methods of warfare. Through the concepts of proportionality and distinction they restrict the application of military force to legitimate military targets and establish that the harm to civilian populations and objects resulting from the use of specific weapons and means of warfare should not be greater than that required in achieving legitimate military objectives. A brief description of the primary treaties limiting military applications of space follows.

The LTBT-1963

The treaty is not concerned with outer space per se, but rather addresses activity in outer space as part of a more general subject, i.e., the prevention of nuclear contamination. Nevertheless, it was the first arms control treaty concerning the legal regulation of the activities of states involved in the exploration and use of outer space. It also was the first legally binding document containing a specific prohibition of the military use of outer space. Under Article I, all parties undertake "to prohibit, to prevent, and not to carry out any nuclear weapon test explosion, or any other nuclear explosion, at any place under its jurisdiction or control: in the atmosphere; beyond its limits, including outer space; or under water." In brief,

- The treaty forbids state parties from carrying out explosion of nuclear devices in the oceans, atmosphere, outer space or in any other environment.
- The treaty bans both nuclear weapon tests as well as peaceful nuclear explosions in outer space.
- The treaty does not contain a verification provision.

The OST-1967[17]

This was opened for signature in 1967 and has been ratified by 97 countries as on 01 January 2005. It constitutes the basic principles governing the behaviour of nations in space. This treaty establishes the basis for all follow-on treaties, thus making it the most important of the space treaties. It is drawn primarily from three UN General Assembly resolutions.[18] It is the world's second non-armament treaty (the first being the Antarctic Treaty), and widely regarded as the 'Magna Carta' of outer space law. It is modelled on the 1959 Antarctic Treaty[19] that prohibited the military exploitation of Antarctica based on the premise that to exclude armaments is easier than to eliminate or control them once they have been introduced. However, unlike the Antarctic Treaty, which prohibits any measures of a military nature, the OST does not ban all military activity.

The articles of the OST constraining military activity are:

- Article I — Outer space, including the Moon and other celestial bodies, is free for use by all states.

- Article II — Outer space, including the Moon and celestial bodies are not subject to national appropriation by claim of sovereignty, use, occupation, or other means.

- Article III — Space activities shall be conducted in accordance with international law, including the UN Charter.

- Article IV — States party to the Treaty undertake not to place in orbit around Earth any objects carrying nuclear weapons or any other kinds of weapons of mass destruction, install such weapons on celestial bodies, or station such weapons in outer space in any other manner.

   - The Moon and other celestial bodies shall be used by all States Party to the Treaty exclusively for peaceful purposes. The establishment of military bases, installations and fortifications, the testing of any type of weapons and the conduct of military manoeuvres on celestial bodies shall be forbidden. The use of military personnel for scientific research or for any other peaceful purposes shall not be prohibited. The use of any equipment or facility necessary for peaceful exploration of the Moon and other celestial bodies shall also not be prohibited.

- Article VII — States are internationally liable for damage to another state (and its citizens) caused by its space objects (including privately owned ones).

- States retain jurisdiction and control over space objects while they are in space or on celestial bodies.
- Article IX — States must conduct international consultations before proceeding with activities that would cause potentially harmful interference with activities of other parties.
  - States must carry out their use and exploration of space in such a way as to avoid harmful contamination of outer space, the moon, and other celestial bodies, as well as to avoid the introduction of extraterrestrial matter that could adversely affect the environment of Earth.
- Article XII — Stations, installations, equipment, and space vehicle on the moon and other celestial bodies are open to inspection by other countries on a basis of reciprocity.

Broadly, the treaty's interpretation implicitly allows the following military activities:

- As per Article III of the OST, space activities are to be conducted in accordance with international law, including The UN Charter, while it would be unlawful for a State to interfere in a hostile manner with the assets in outer space of another State (Article-2 of UN Charter), conversely a state could also legally use military force to defend itself against hostile actions (Article 51 of UN Charter).
- Paragraph 1 of Article IV implies that objects carrying nuclear weapons or other Weapons of Mass Destruction (WMD) can freely transit outer space,[20] as long as they do not "orbit" the earth. Likewise, WMD[21] that escape Earth orbit are permitted, except that they may not be "installed" on celestial bodies or otherwise "stationed" in outer space. Other 'non-nuclear/non-WMD' weapons may be placed in orbit (but not on the moon or other celestial bodies) and used to attack targets in space or on the earth. Hence, Kinetic Energy Weapons, Directed Energy Weapons, armed re-usable space planes,[22] and such like systems are permitted.
- Secondly, para 1 of the above treaty refers only to "celestial bodies" and "outer space" and not to "outer space, the moon, and other celestial bodies," as in other provisions of the treaty, allowing

scope for the question (largely academic) as to whether weapons of mass destruction are expressly banned (or not) from the Moon, as well as from trajectories to and around it. Conversely, para 2 refers only to "the Moon" and "other celestial bodies" which implies that while WMD may be tested in outer space, the same cannot be tested on the moon. On similar lines, it also implies that countries may create military bases, installations and fortifications in outer space (e.g., on orbiting satellites like the ISS), though not on the moon or other celestial bodies. They may use satellites to perform all manner of military functions, including communications, reconnaissance, navigation, geodesy, etc.

- Nuclear-powered satellites are permitted.
- There is no direct ban on non-nuclear anti-satellites or anti-missile weapons, whether space or Earth-based.

In view of the foregoing it is apparent that the OST is largely true to generic concepts and ideas and open to very broad interpretations. The OST has been overtaken by technology and its own limitations. Whilst the rules system developed by the OST is fairly comprehensive, it does not apply to the generation of space weapons currently being considered.[23] This in turn implies that new generations of space weapons could be developed and deployed without abrogating the OST.

The Moon Treaty, 1979[24]

This treaty reiterates for the moon many of the principles found in the OST. The Agreement's language pertaining to military usage largely mirrors Article IV of the OST. For example, Article III prohibits the threat or use of force or any other hostile act on the moon and the use of the moon to commit such an act in relation to the earth or to space objects. However, regarding military activity, the Agreement forbids the placement of weapons of mass destruction, including nuclear weapons, on the moon itself, in orbit around the moon, or on trajectories to and around the moon, and on other celestial bodies. Further, the Agreement's military provisions do not prohibit the placement of weapons in outer space in general, only weapons of mass destruction.

By 1984, the Moon Treaty had been ratified by five states and in accordance with its provisions had entered into force; however, although the Treaty was negotiated by consensus, it had not been ratified by the major space powers, and even now (as on 01 January 2005) after 25 years of its inception, it enjoys a ratification by only eleven nations, none of whom is a space-faring nation or for that matter a space-launching power possessing the requisite technical means to launch objects into outer space and to explore and exploit the resources of the moon.

EnMod[25]

The 1976 Convention on the Prohibition of Military or Any Other Hostile Use of Environmental Modification Techniques forbids State Parties to:

Art. I.1. *"engage in military or any other hostile use of environmental modification techniques having widespread, long-lasting or severe effects as the means of destruction, damage or injury to any other State Party"*.

Article II has parties which agree not to engage in military or hostile environmental modification activities, defined as *"any technique for changing—through the deliberate manipulation of natural processes— the dynamics, composition or structure of the Earth, including its biota, lithosphere, hydrosphere and atmosphere, or of outer space."*

Art. III.1 states that *"The provisions of this Convention shall not hinder the use of environment modification techniques for peaceful purposes (...)"*.

One important weakness of this treaty is that it does not prohibit all major activities that could negatively affect the earth's environment. Non-military EnMod experiments could be as damaging as those intended for military or hostile purposes. For example, the "Thunderstorm" Solar Powered Satellite (TSPS)[26] envisaged for peaceful weather modification to prevent formation of tornadoes would be absolutely legitimate since it would be within the aegis of the above-mentioned Article III.1. However, as and when conceived, the system would have immense potential for military misuse and such a possibility cannot be simply wished away.

International Telecommunication Union[27]

The ITU can trace its official existence to the International Telegraph Convention of 1865. From its humble beginnings in the 19th century, the ITU has grown into an immense organisation with a voting membership that very nearly mirrors that of its parent organisation, the United Nations. The United Nations has tasked the ITU with the responsibility of coordinating allocation of GEO orbital positions and operating frequencies. The ITU, under the auspices of the United Nations, has been coordinating space radio communications and managing satellite spectrum and slot allocation since 1963. The ITU has, by design, no enforcement powers or authority, member states choose voluntarily to abide by ITU rules and regulations and work to resolve any conflicts about spectrum usage in good faith. The authority to place a satellite into orbit and employ frequencies for its use rests with each sovereign state. ITU is intended to be a coordinating agency, which through its Radio Regulations Board (RRB) coordinates the international use of the radio spectrum. This coordination and regulation of orderly exploitation of the EM spectrum is designed to avoid harmful radio frequency interference, potential disputes and conflicts.[28]

All nations are supposed to register with the ITU for both position and frequency before launching and operating a satellite in GEO. Once an orbital position and frequencies have been registered, the registration remains in effect until the operator-defined system life expectancy has expired, or until the ITU is notified that the frequency and orbital position are no longer in use by the registrant. When disputes arise, they are managed on an ad hoc basis. The parties can resort to any of a wide variety of techniques, including negotiation, settlement in accordance with non-ITU dispute resolution agreements that may be binding on the parties, or any other mechanism agreed upon by the parties.

Although not applicable to the military or other national security functions since the ITU has no jurisdiction over the use of the spectrum for military purposes, the ITU regulations govern the majority of telecommunications systems in space. During military operations, and especially during armed conflict, the military must operate its telecommunication networks, or lease the capability from civilian providers, so as to avoid radio interference. Since the armed forces rely heavily on telecommunications for efficient command and control, including commercially operated telecommunications systems, their use of the radio spectrum must be done taking into account other users with the potential for harmful interference. Here, it would be pertinent to note that the 1932 ITU convention as amended in 1992, 1994 and 2002 also protects civilian satellites from interference.

Like most international legal regimes, the ITU does not have the authority to enforce its ruling directly; it is dependent upon its own legitimacy to influence the behaviour of its members. The ITU has no policing powers, but member-nations generally abide by the regulations, motivated both by international treaties and consideration of mutual benefit. Nevertheless, member-nations can also deviate from the rules in matters of national defence.

Missile Technology Control Regime

This is a consensual protocol signed by the 33 nations, including most of the world's space-faring nations to restrict exports of missile-related technologies to countries that are not members. Several other States (including China and Israel) have pledged to adhere to the MCTR without formally joining the Regime. India is not party to the MTCR. The stated goal of the MTCR, which was established in 1987, is to restrict the proliferation of missiles capable of carrying WMD. Space launch vehicles are considered "missiles" and are therefore covered by the MTCR. The MTCR is a voluntary arrangement, not a formal international agreement. Accordingly, each Partner State implements the Regime on a national level through its own national export control regulations. There are no provisions in the regime for enforcement of its terms or sanctions for violations.

However, ballistic missiles are not the only potentially harmful application of dual-use space technology. For example, space platforms for scientific or commercial uses could be converted into orbital weapons carriers, and any object in space with sufficient manoeuvrability can be used as a kinetic energy weapon. It is unlikely that the MTCR is comprehensive enough to confront future contingencies of this nature. Another factor is the exclusive character of the regime, a quality that could weaken its effectiveness in the future. While the MTCR has achieved limited success in stemming the spread of missile technology, it has also had the effect of inhibiting the development of certain nations' peaceful space programmes.

Figure 5.3 attempts to encapsulate the major legalities impacting military space activities.

Major Legalities Impacting Military Space Activities

Legislation	Principle/Constraint	Remarks
UN CHARTER (1947)	• Article 2(4) of the UN Charter prohibits the threat or use of force in international relations. • Article 51 codifies the right of self-defence in cases of aggression involving the illegal use of force by another state(s).	Made applicable by Article-III of the OST.
LTBT (1963)	• Bans nuclear testing and nuclear explosions (both peaceful & otherwise) in space.	Covers lacunae of para 2 Article IV of OST which implies that WMD may be tested in space but not on the moon.
OST (1967)	• Bans deployment of nuclear weapons & WMD in space.	Allows deployment of non-nuclear, non-WMD, conventional & other weapons including non-nuclear ASATs and other satellites performing passive military functions like ISR, communication, navigation, etc.
OST (1967)	• Bans creation of military bases, installations and fortification on celestial bodies.	Permits creation of military bases like space-stations in orbit.
OST (1967)	• Bans testing of any type of weapons on celestial bodies.	Permits testing of conventional weapons in space.
OST (1967)	• Bans conduct of military manoeuvres in space.	—
OST (1967)	• Directs states to conduct international consultations before proceeding with activities that would cause potentially harmful interference with activities of other parties.	Effectively contains peace-time jamming, spoofing, disruption, etc. However, the above are permitted during hostilities or for self defence.
OST (1967)	• Bans apportioning of space or celestial bodies.	Ban on space resources is ambiguous since spatial locations, frequency spectrum, solar power,[29] etc. are already in military use. However, no such ambiguity or debate on apportioning celestial bodies exists.
Registration convention	• Ban on launching space objects without notifying the UN.	Hinders covert military activity in space.
Astronaut rescue agreement	• Bans hindering the rescue and return of astronauts & space objects.	Inspection of space objects before return to launching party not prohibited.
EnMod	• Bans engagement in military or any other hostile use of environmental modification techniques having widespread, long lasting or severe effects as the means of destruction, damage or injury to any other State Party.	Non-military EnMod activities could be as damaging as military EnMod activities.

Figure 5.3

Legal Complexities of "Dual-Use" Space Assets

In addition to the above, the complexities arising out of the "dual-use" characteristics of space assets are enormous. It is a well-known fact that most military "force-enhancement" missions like communication, navigation, remote sensing, etc. could be carried out by dual-use satellites instead of dedicated military satellites. Dual-use complicates matters relating to isolation of the legal status of satellites considering that a satellite may have dual-use both in terms of utility as well as ownership. For example, in terms of utility, space launch vehicles could be used as ballistic missiles, civilian communication satellites can carry military communication transponders (and vice-versa), military navigation satellite constellations have immense civilian applications, remote-sensing satellites could be used for both military as well as agricultural purposes. The ownership of the satellite can be dual as between several states directly, or through participation in intergovernmental organisations such as the International Telecommunications Satellite Organisation (INTELSAT) and the former International Mobile Satellite Organisation (INMARSAT).[30] When ownership of a space asset is shared among several States or entities, the legal status of parties as belligerents, neutrals, etc. further obfuscates and complicates the application of legalities as in case of the ARABSAT telecommunication system whose transponders were put to military use by both the coalition as well as Iraqi forces.[31]

Legal Implications of the Military Use of Civilian Systems

The Law of Armed Conflict (LOAC, also called the "law of war") is the branch of international law regulating armed hostilities. As armed forces and civilian users increasingly depend on the same commercial space systems, the application of LOAC principles is becoming more complicated. Moreover, the fact that in many cases civilian control systems vital to militaries during times of armed conflict raises certain ethical and practical issues that cannot be ignored. Under LOAC principles, legitimate military targets must be distinguished from protected civilian objects. Anticipated collateral damage must be weighed against expected military advantage, and excessive civilian damage avoided. However, force may lawfully be used against objects which an adversary is using for a military purpose, if neutralisation of the object would offer a definite military advantage. The analysis becomes more complex, however, when the object being used by the adversary belongs to a "neutral" third party.

Non participants in a conflict may declare themselves to be neutral. As long as the neutral State does not assist either belligerent party, it is immune from attack by the belligerents. However, if one of the belligerents uses the territory of a neutral nation in a manner that gives it a military advantage and the neutral nation is usable or unwilling to terminate this use, the disadvantaged belligerent has the right to attack its enemy in the neutral state's territory. Hence, if a neutral State permits its space system to be used by a belligerent military, the opposing belligerent would have the right to demand that the neutral State stop doing so. If the neutral State is unwilling or unable to prevent such use by one belligerent, it would seem reasonable to authorise the other belligerent to prevent the offending use. In the context of space systems used in time of conflict, before resorting to force a belligerent could (or should) demand a neutral nation not provide satellite imagery, navigation services, or weather information to its adversary.[32]

Status of Civilians Supporting Military Space Activities

A corollary to the problem of armed forces and civilian users relying on the same space systems is the increasing use of civilians to perform military jobs. In addition, civilian systems are providing certain information and services formerly provided by military systems. In space, this trend is especially noticeable in high-tech fields such as satellite control, ground systems maintenance, and satellite data-collection and interpretation. However, the LOAC requires a distinction to be made between combatants and non combatants. Only combatants who are members of a State's armed forces have the right to participate directly in armed conflict. The term non-combatant is generally synonymous with civilian. Civilians are not authorised to take a "direct part in the hostilities". Persons who commit combatant acts without authorisation are unlawful combatants and are subject to criminal prosecution. If combatant acts are conducted by unauthorised persons, their national government may be in violation of the LOAC. Thus, the participation of civilians in the context of space operations supporting armed conflicts presents difficult and complex issues.[33]

The law of war has traditionally recognised that civilians may participate in a war effort without being declared unlawful combatants. However, acts intended or likely to cause actual harm to enemy armed forces are prohibited by non combatants. It is therefore generally agreed that non combatant participation in activities such as weapons production, military engineering, and military troop transport is not prohibited, even though these acts ultimately harm an enemy. However, there exists no such general agreement about whether the gathering and dissemination of intelligence and the transportation of weapons is direct participation or not.

Definitional Issues and Other Controversies

Most international legislation on space was developed in the aftermath of the 'Sputnik era' and was quite comprehensive for its times. However, the fact that a lot of space legislation is now mired in controversies over definitional issues and self-suited interpretations indicates that while the legislation was comprehensive enough for the technological and other challenges of those times, it certainly has been unable to evolve sufficiently to contain the technological and other challenges of the present. Secondly, there are several other issues relating to the wording of the space treaties themselves that deserve closer analysis but are beyond the scope of this paper. Nevertheless, some of the major controversial issues have been dwelt upon briefly, as under.

The term "peaceful"

In spite of the apparent ubiquity of this term in various UN documents and space treaties, it continues to lack a precise authoritative definition and hence is open to self-serving interpretations. The initial and widespread interpretation of the term in relation to space was "non-military" and the same was acceptable to most nations including the US and the USSR. However, soon after the launch of its early satellites, the US began changing its position claiming that the term meant "non-aggressive" and not "non-military". The Soviets initially held on to the latter interpretation, but eventually accepted the former interpretation. By this time both had in orbit satellites conducting military operations of Early Warning, surveillance, etc. and the term soon became synonymous globally with "non-aggressive". The interpretation of the term continues to expand as per state interests and practice. Whereas the results of the attempts in interpretation remain unfinished to this day as per the 1969 *Vienna Convention on the Law of Treaties,* the words in a treaty must be interpreted in accordance with their ordinary meaning.[34] In general the term peaceful is defined as disposed or inclined to peace: aiming at or making for peace; friendly, amicable, pacific. It is obvious that this description cannot be applied to any current or past military use of space.

Delimitation of "Outer-Space"

The Concise Oxford Dictionary (9th edition) defines Outer Space as "the physical universe beyond the earth's atmosphere"; however, there exists no consensus definition in the UN on the term. The question on where the earth's atmosphere ends (in which states retain sovereignty) and where the physical universe beyond begins (in which concepts of sovereignty are not applicable) remains unanswered. Discussions regarding delimitation of outer space occurred as early as the 1930s. Yet there is still no satisfactory legal definition of the same.[35] The scope of applicability of numerous principles of space law like sovereignty, appropriation, etc. is constrained on account of the lack of a precise definition of the term and a lack of agreement on a precise boundary delimiting air and space. Nevertheless, states like Australia have enacted legislation to the effect that launches from Australia would be licensed under the Australia Space Activities Act if the launch vehicle and/or payload is intended to reach an altitude of at least 100 km above sea level.

Ambassadorial Status of Astronauts

Article V of the OST bestows on Astronauts the unique Ambassadorial status of "envoys of all mankind" and enjoins that certain privileges be afforded to such envoys. This is a vestige of the pioneer era when Astronauts/Cosmonauts were personnel of select calibre representing particular states. However, with space tourism in the offing a variety of passengers is imaginable who would be driven by the quest of pleasure and self-indulgence rather than national pride, scientific pursuit, social welfare, public development or any such lofty motive. At the cost of sounding rhetorical it would be interesting to imagine a hypothetical scenario wherein Saudi multimillionaire Osama Bin Laden manages a "space tour". Whether the same lofty Ambassadorial status would be bestowed on him or not is an issue worth mulling over.

Space Legislation on Information Warfare (IW)

It is a well-known fact that space-based information and consequent IW capabilities are the driving force of the Revolution in Military Affairs (RMA) which is presently the most sought-after capability of most nations. However, none of the prevailing legislation space expressly prohibits IW activities that make use of space-borne assets. In fact, no provision of International Law expressly prohibits what is known as IW. Secondly, satellites since the beginning of the space-age have been routinely used for IW activities like military communication, navigation, ISR, etc. This has been facilitated in no small measure by the loose-ended interpretation of the term 'peaceful'. IW activities, on account of their non-lethal, non-intrusive character are considered to be within the realm of 'non-aggressive' and hence legitimate.

Legislation Pertaining to Military Remote Sensing

It is common knowledge that high-resolution imaging has the potential to upset delicate balances of power or aggravate existing asymmetric in military capability.[36] As regards remote sensing, at the international level, the UN General Assembly Resolution on the Principles of Remote Sensing does not directly address national security concerns. The Principles are of a generic character and do not directly address the military aspects of remote sensing. In fact the principles are largely silent with regard to military remote sensing which in effect means that it does not place any meaningful restraints on the use of remote sensing technology by the military. Also, the principles are not regarded as legally binding. Nevertheless, certain nations do subject commercial satellites to domestic restrictions to address national security concerns.

Sale of Real Estate on the Moon

In recent times a lot of controversy has been generated on "non-issues" like sale of real estate on the moon and other celestial bodies. Most of it is attributed to certain individuals and organisations attempting to exploit the clause that Article II of the OST refers only to "national appropriation" and not "private appropriation". However, the OST only permits states to exercise jurisdiction over their space objects, personnel, and arguably, zones of safety in the vicinity of their space objects and in areas where their citizens are conducting ongoing, significant activities. States have no jurisdiction or authority over any other areas of outer space or celestial bodies, and therefore cannot grant or recognise permanent, immovable property rights in those areas. Granting or recognising such rights beyond the limits of national jurisdiction, and thereby subjecting said areas to national jurisdiction, would violate Article II of the OST because it would constitute "national appropriation...by any other means".[37]

The above anomalies are in no way exhaustive; in addition to the above, certain issues like liabilities, the differences between "appropriate" and "launching states" as well as the definition of "appropriate state" have also not been clearly defined but are presently not discussed in view of the compulsions of brevity.

Conclusion

In view of the foregoing, it is apparent that the prevailing legislation regarding military space activities is inadequate and incomplete. The legislation requires to be made more comprehensive in the light of the technological and other challenges of the new millennium. Nevertheless, considering the fact that a certain amount of legislation on the subject is already in place, it can safely be assumed that the same would form the basis for evolution and enactment of further legislation regarding military space activities. Thus, apart from a few arenas which may experience drastic departures from past legislation, most of the legislation would build upon the existing premises and principles. In this context, it would be imperative to first comprehensively assess the implications of prevailing legalities and then contemplate building of military space capabilities. This would ensure the growth of capabilities within the realm of prevailing legalities and also endow flexibility for expansion as and when the eyes of law are opened by experience.

Notes and References

1. The Soviets were the early pioneers of space law. Dr. Stephen E. Doyle in his book *Origins of International Space Law* refers to papers presented at an air law conference in Moscow in 1926 which deals with debates about the nature and extent of national sovereignty in airspace. Additionally, mention is also made of Czech professor Vladmir Mandall writing the world's first major work of substance on the subject of space law in 1932.

2. General Assembly resolution 1148 (XII) of 14 November 1957.

3. Refer <http://www.oosa.unvienna.org/SpaceLaw/gares/html/gares_13_ 1348.html>.

4. See "Space Law: Frequently Asked Questions" at site of UNOOSA, www.oosa.unvienna.org/FAQ/splawfaq.htm

5. Ibid.

6. For a more detailed description on space-related treaties, agreements and resolutions and resolutions, visit the website of UN Office of Outer Space Affairs at www.oosa.unvienna.org/SpaceLaw/spacelaw.htm.

7. Refer to the seventh question of "Space Law: FAQ" at site of UN Office of Outer Space Affairs at www.oosa.unvienna.org/FAQ/splawfaq.htm

8. For example, in July 2002, the Chinese Embassy, the Xinhua News Agency, etc. alleged that the Chinese extremist cult "Falungoing" attacked SINOSAT2A and SNOSAT3A and hijacked nine China Central Television Station (CCTV) channels, 10 provincial TV channels and was known to have interfered with Chinese communications satellites by penetrating the broadcast signals of Chinese communications satellites. In at least one case, a Chinese spacecraft suddenly started transmitting Falungong programming; in another, the cut blocked the signal totally. See details at www.china-embassy.org/eng/zt/ppflg/ t36611.htm.

9. Commercial Space Launch Act of 1984, 49 U.S.C. [section] 701 (1994 as amended in 1998).

10. Ref Para 24 (d) of No 104—Statute on Licensing Space Operations of Russian Space Law at www.oosa.unvienna.org/SpaceLaw/national/russian_federation/ decree_104_1996E.html

11. Ref Ch. 38, Para 4(2) of Outer Space Act (United Kingdom, 1986) at www.oosa.unvienna.org/SpaceLaw/national/united_kingdom/ outer_space_act1986E.html

12. Space Activities Act of 1998, Acts of Parliament of the Commonwealth of Australia No. 23, s. 18(e) (assented to Dec. 21, 1998).

13. Space Affairs Act, Statutes of the Republic of South Africa No. 84 of 1993, [section] 11(2) (assented to Jun. 23, 1993 (commenced 6 Sep. 1993).

14. Space-faring nations are generally considered as those possessing the full complement of space capabilities including indigenous space launch, satellite manufacture and allied ground infrastructure.

15. The above is the status as on 01 Jan 2005, previous chart on space treaties, principles and agreements expands on the abbreviations of OST, BRS, ITSO, etc. In case of the above, R is indicative of ratification, S indicates signature and D indicates declaration of acceptance of rights and obligations. European efforts being a joint endeavour, nations are not treated separately.

16. C. Jayaraj, "India's Space Policy and Institutions", paper presented at United Nations/Republic of Korea Workshop on Space Law: "United Nations treaties

on outer space: actions at the national level" Daejeon, Republic of Korea, 3–6 November 2003.

17. Until 13 June 2002, when it expired, the ABM Treaty of 1972 banned space-based missile defence systems between the US and the USSR. Both SALT-1 and SALT-2, although primarily about arms control on land, also had an outer space component. SALT-1 (along with the ABM Treaty) prohibited interference with national technical means of verification (i.e. satellites). The SALT-2 agreement prohibited the development, testing or deployment of weapons of mass destruction in space. The 1991 Strategic Arms Reduction Treaty I(START-1) also overlapped with the OST's prohibition on the placement of weapons of mass destruction in outer space. START-1 had other potential implications for military space operations as well. There are several restrictions on the use of Intercontinental Ballistic Missile (ICBMs) or Submarine Launched Ballistic Missile (SLBMs) as space launch boosters.

18. For a brief of overview of the basic principles of the treaty, see site of OOSA at www.oosa.unvieinna.org/SpaceLaw/outerspt.html and for overview of the articles refer to Appendix-B.

19. The resolutions include: (a) *Resolution 1772, Jan. 3, 1962,* International Co-operation in the Peaceful Uses of Outer Space; (b) *Resolution 1962 (XVIII), Dec. 13, 1963,* Declaration of Legal Principles Governing Activities of States in the Exploration and Use of Outer Space, and (c) *Resolution 1963 (XVIII), Dec. 13, 1963,* International Co-operation in the Peaceful Uses of Outer Space.

20. For details of the Antarctic Treaty, visit www.antarctica.ac.uk/About_Antarctica/Treaty/treaty.html

21. The provision was deliberately worded to permit the earthly use of ICBMs, which transit through space, due to the fact that ICBMs already included the arsenal of both the Space powers of US and USSR as well as their national defence systems which were both based upon ICBMs.

22. Weapons of mass destruction were defined in 1948 by the UN Commission for Conventional Armaments as *"those which include atomic explosive weapons, radioactive material weapons, lethal chemical and biological weapons, and any weapons developed in the future which have characteristics comparable in destructive effect to those of the atomic bomb or other weapons mentioned above" [UN document S/C.3/32/Rev.1, August 1948].* However, the above needs to be reconsidered in view of UN General Assembly Resolution 51/37 of 7 January 1997 [A/RES/51/37] in which it expresses its determination *"to prevent the emergence of new types of weapons of mass destruction that have characteristics comparable in destructive effect to those of weapons of mass destruction identified in the definition of weapons of mass destruction adopted by the United Nations in 1948" and it "reaffirms that effective measures should be taken to prevent the emergence of new types of weapons of mass destruction."* Thus it can safely be surmised that there exists no precise, exclusive definition of WMD, which in turn further

opens the treaty to interpretations suited to individual requirements and allowance for more lethal non-WMD like HPM, CAV, etc.

23. Aerospace planes can hence operate in space under modest legal restrictions. Problems arise only due to national sovereignty in air space.

24. For a detailed description of such weaponry developed or under consideration which circumvents the OST, refer *Yorkshire CND brief* of June 2004, "Fighting for Space" at http://cndyorks.gn.apc.org/yspace/overview.htm

25. For a detailed brief of the agreement and the articles contained therein, see www.oosa.unvienna.org/SpaceLaw/moontxt.html

26. For a detailed brief on EnMod and the articles contained therein, see www.fas.org/nuke/control/enmod/text/environ2.htm

27. Eastlund Bernard J., "Mesocyclone Diagnostic Requirements For Thunderstorm Solar Power SatelliteConcept" *Proceedings of the Second Conference on the Applications of Remote Sensing and GIS for Disaster Management,* 19–21 January 1999, p. 1.

28. For further details on ITU, visit its site at http://www.itu.int/home/

29. The potential for conflict on account of orbital crowding in the geostationary orbit slots over Asia is enormous considering that in 1994, China's placing of its APSTAR1 communication satellite in an unauthorised GEO slot with only one degree of separation between satellites operated by Japan and an international consortium in a slot "owned" by Tonga brought it into conflict with Japan. After tremendous pressure was brought on the Chinese, they finally agreed, after six weeks, to move the satellite to another slot which they had to lease from Tonga. In 1996, severe crowding in the geostationary orbital slots over Asia led to the jamming of a communication satellite by PT Pasifik Satellite Nusantara (PSN) of Jakarta, Indonesia, in defence of an orbital position claimed by Indonesia. The ITU did not get involved in the dispute settlement process, claiming bilateral negotiations were appropriate. The PSN satellite project was later shelved on account of a monetary crisis rather than negotiation or dispute settlement and finally though Indonesia has a legitimate claim to the position, China continues occupation of this position.

30. Debate and dispute on whether solar energy is a spatial resource (as it is immaterial. like the frequency spectrum) or a celestial body resources (as it originates in the Sun) and thus whether it can be consecrated for peaceful purposes only continues unabated. See Virgiliu Pop, "Security Implications of Non-Terrestrial Resource Exploitation", IISL-00-IISL.4.11. Paper presented at the 43rd Colloquium on the Law of Outer Space, 51st International Astronautical Congress, Rio de Janeiro, 6 October 2000, Proceedings of the 43rd Colloquium on the Law of Outer Space, pp. 335–345 at http://www.spacefuture.com/archive/security_implications_of_non_terrestrial_resource_exploitation.shtml

31. Now privatised, INMARSAT has been used in support of several previous armed conflicts though its use among coalition forces during Operation Desert Storm in 1991 was the most publicised.

32. F.R. Cleminson, "Banning the Stationing of Weapons in Space Through Arms Control: A Major Step in the Promotion of Strategic Stability in the 21st Century", in *Arms Control and the Rule of Law: A Framework for Peace and Security in Outer Space* 39 (J.M. Beier & S. Mataija, eds., 1998).

33. Elizabeth Seebode Waldrop, "Integration of military and civilian space assets: legal and national security implications", *Air Force Law Review, Spring 2004* available at www.findarticles.com/

34. Ibid.

35. The above was suggested to the author by Ms Sama Payman, Legal Officer, Committee Services and Research Section, United Nations Office of Outer Space Affairs.

36. The atmosphere has no abrupt cut-off. It slowly becomes thinner and fades away into space. There is no definite boundary between the atmosphere and outer space. Three-quarters of the atmosphere's mass is within 11 km of the planetary surface. An altitude of 75 miles, 120 km (400,000 ft) marks the boundary where atmospheric effects become noticeable during re-entry. In general parlance, the Karman line, at 62 miles (100 km), is also frequently used as the boundary between atmosphere and space. In the United States, persons who travel above an altitude of 50.0 miles (80.5 km) are designated as astronauts.

36. For a more detailed brief on military remote sensing, refer to Vipin Gupta, "Offensive Military Application of High-Resolution Imaging" at www.llnl.gov/csts/publications/gupta/risks.html

37. For an exhaustive discussion on the above refer to Wayne N. White Jr. "Interpreting Article II of the Outer Space Treaty at http://www.spacefuture.com/archive/real_property_rights_in_outer_space.shtml

Chapter 6

The Great Asian Space
Militarisation Race

CHINA'S MILITARY SPACE PROGRAMME

The two Asian giants, India and China are ancient civilisations bearing their own distinctive cultural, historical, intellectual baggage and survival skills. Thus, while the Chinese pioneered the development of Chemistry, suspension bridges, etc. the Indians discovered zero, finger-printing, etc. Similarly, while the Chinese are known to have invented the rocket in 970, it was the Indians who used them decisively for the first time in battle. These analogies are indicative of the fact that both possess a civilisational culture of intellectual and scientific pursuit in spite of repetitive threats to survival amidst hostile environs and invasions. Hence, while the astounding progress by both countries with regard to scientific endeavours in space is not surprising, the overwhelming emphasis of the one on space militarisation and the near-total disregard by the other of militarisation or for that matter even space security (securing its assets in space) is truly amazing.

Notwithstanding Chinese public proclamations on peaceful uses of outer space, the fact is that the Chinese space programme especially in military terms is evolving at an extremely rapid pace and apart from out-racing every other worthwhile space power in the Asian region, has now out-distanced every other Asian space power to the extent that it has decisively altered the 'balance of power' overwhelmingly in its favour and is likely to tilt the scales further in the next few years. Such a perceptible imbalance, apart from causing general consternation within Asia may cause an 'action–reaction' effect reminiscent of the nuclear order. For example, the American use of space for military purposes was in response to the Soviet threat; the Chinese claim theirs is in response to America's; the Indians and Japanese might be triggered off by Chinese ambitions, and so on and so forth setting off a chain reaction across the continent, with risks of planning on worse-case scenarios. Therefore, an objective assessment of China's military space programme has been undertaken, which is as given below.[1]

Dual Use Capabilities

Apart from the fact that China's space programme is one of the least publicised in the world is the aspect that it has largely succeeded in shrouding its military space programmes under a cloak of secrecy and civilian nomenclatures. The nature of space technology and the overlapping characteristics of the Chinese military and civil space programme permitted the parallel development of a missile programme and a space launch vehicle. From its beginning in the 1950s, the Chinese have adapted their ballistic missile programme into major space programmes. Its space programme was primarily an offshoot of its ballistic missile programme. The development of ballistic missile technology had initially triggered Chinese interest and growth in space. In the early years the Chinese missile programmes were given priority and Chinese space assets came about as derivatives of these projects. Early Chinese space endeavours were based upon its ballistic missile rocket technology. Once China achieved proficiency in space launch technology, it diversified to develop its satellite applications, particularly in a military context. Much of China's space programmes are deemed to be civilian, but have dual use, especially with regard to military capabilities. These developments were initially meant to fulfill the 'force-enhancement' missions of surveillance and reconnaissance, communications, navigation, ELINT (Electronic Intelligence), etc. and following a certain amount of maturisation in these roles, the Chinese have gradually moved on to dedicated satellites for military purposes as well as 'space-control' mission programmes by developing ASAT (Anti-Satellite), space-based ISR (Intelligence, Surveillance, Reconnaissance) and other counter-space developments. Thus from its very beginning, the conceptualisation, design, and evolution of China's space programme has always had a pronounced military orientation. Consequently, its overall control has always rested with the Central Military Commission.

Broad Strategic Drivers

While the potential military utility of space systems was at the heart of China's decision in 1963 to undertake its own space programme, the centrality of space technology in bolstering the Revolution in Military Affairs (RMA) and by extension overall military force capabilities was comprehended significantly after Gulf War 1. Driven largely by the Chinese Academy of Military Sciences (CAMS) the Chinese meticulously studied the tremendous force-multiplication "effects" enabled by space during the 1991 Gulf War and the recent conflicts in Kosovo, Afghanistan and Iraq and reached the conclusion that space power is an essential element of effective military action. In particular, the tremendous contribution of space to RMA[2] and hence modern high-technology warfare was well understood by Chinese analysts. Thereby a strategic re-think of military concepts and doctrine was carried out, leading to traditional concepts being replaced with modern warfare strategies and space-enabled techniques. It was such comprehension that led to the modification of the guiding principles for PLA modernisation from "local, limited war" to "limited war under high-tech conditions". Most Western analysts are of the opinion that China understands its slim chances against technologically superior opponents like the US, hence the desire to have some chances of success against a technologically superior opponent drives China to investigate inherently riskier asymmetrical advantages. Notwithstanding Chinese and Western perspectives on the subject, the bottom line is that China doctrinally comprehends the tremendous impact of space on conventional capabilities and has earnestly begun the pursuit of 'operationalising' space-enabled capabilities.[3]

Launch Capabilities

Even as Beijing publicly declares that space should not be militarised and that space technologies should be used for peaceful purposes, military considerations play an important role in Chinese space programmes, owing in part to the programme's military beginnings. For instance, China's three stage CZ-1 (*Chang Zheng*-1/Long March-1) SLV (Space Launch Vehicle) is a derivative of the military's DF-4 (*Dong-Feng*/East Wind-4) ballistic missile. Likewise, versions of the DF-5 have also become SLVs, specifically the FB-1 (*Feng Bao*/Storm-1) and the contemporary Long March (CZ-2) used to launch satellites and the *ShenZhou* series spacecraft.

In fact, the technologies used for Long March and Chinese ICBMs (Inter Continental Ballistic Missiles) were so similar that in 1998 the US Senate select committee on intelligence warned that technical assistance once provided by American companies to improve China's Long March rockets may have inadvertently threatened US national security by improving the accuracy and reliability of Chinese ICBMs targeting the US. The Chinese firm that launches Long March, the China Great Wall Industry Company, has been sanctioned by the US government for missile proliferation.[4]

It must also be borne in mind that though the CALT (Chinese Academy of Launch Vehicle Technology) under CASC (China Aerospace Science and technology Corporation) has general responsibility for manufacture and design of launch vehicles and analogous ballistic missiles, it is finally the PLA (People's Liberation Army), and within the PLA, the GAD (General Armament Department) which controls all space launch and tracking facilities. Additionally, the Second Artillery Corp—China's nuclear strike force—is also one of the primary drivers for China's launch vehicle programmes and focuses on integrating its missile forces into multiservice operations. Secondly, China already has a ballistic missile modernisation programme under way to upgrade all classes of missiles both qualitatively and quantitatively. China is known to have a deployed inventory of about 500 SRBMs. China is replacing all of its roughly 20 CSS-4 Mod1 ICBMs with longer range CSS-4 Mod2. Development of the DF-31 ICBM is progressing, and deployment is expected to begin later this decade. China also is developing two follow-on extended range versions of the DF-31: a solid propellant, mobile ICBM and a solid-propellant submarine-launched ballistic missile, the JL-2. The Second Artillery is continuing to supplement its aging inventory of liquid propellant CSS-2 IRBMs with solid-propellant, road-mobile MRBM.[5] Among China's new ICBMs, the DF-31 has a civilian derivative for space launches, the *Kaitozhe*-1 (Pioneer-1). As the *Kaitozhe* is further developed, the DF-31 will be improved, as well. China's ballistic missile programme is known to be largely Taiwan and US-centric which is of little consolation to India, in view of the highly mobile character of ballistic missiles. China is expected to increase the lethality and accuracy of its 500-strong SRBM force through use of satellite-aided guidance systems. In view of the homogenous nature of its space, ballistic missile and nuclear programme, it can be safely surmised that China has a fairly advanced base for doctrinally and operationally integrating its capabilities to not only gain an "asymmetric advantage" against super-powers like the US but also to strengthen its dominating posture against lesser rivals within Asia. Details on Chinese ballistic missiles are given in Figure 6.1.

Details of Chinese Ballistic Missiles

Delivery Vehicle		Type	Range (km)	Estimated Numbers[6]		
Chinese Designation	Western Designation			SIPRI (2004)	Mil Balance (2003-04)	CDI Weapons database (2002)
DF-3A	CSS-2	IRBM	2,800	40	32	40
DF-4	CSS-3	IRBM	5,500	12	20	20
DF-5/5A	CSS-4	ICBM	13,000	20	24	20
DF-21A	CSS-5	MRBM	1,800	48	60	48
DF-15/M-9	CSS-6	SRBM	600	–	24	–
DF-11/M-11	CSS-7	SRBM	300	–	32	–
DF-7/M-7	CSS-8	SRBM	160	–	30	–
DF-31	CSS-9	ICBM	8,000	–	08	–
JL-1	CSS-N-3	SLBM	1,700	12	12	12

Figure 6.1

Space System Capabilities and Programmes

The dual-use potential of the Chinese space programme is not limited to civilian rocketry and military missiles. Broadly speaking, the Chinese space programme objectives by 2010 include creating an integrated military and civilian earth observation system; building a Chinese-operated satellite broadcasting and telecommunication system to be used for both civilian and military purposes which finally would be used to link the Chinese military forces; establishing a Chinese run GPS (Global Positioning System) and upgrading China's Long March rocket, while continuing to develop a low-cost successor.

Signals Intelligence (SIGINT)

However, narrowing down to specifics, with regard to SIGINT and its two important subsets of ELINT (ELectronic INTelligence) and COMINT (COMmunications INTelligence), it first needs to be borne in mind that China has a ground-based SIGINT network that is widely regarded to be the most extensive in all Asia. This capability is based on ground stations, ships, aircrafts and other mobile platforms. The entire programme is managed by the Third Department of the General Staff Headquarters, responsible for monitoring foreign communications and producing finished intelligence. To complement and augment the above capability, China's first ELINT satellites, the JSSW were developed in the early 1970s under its 701 programme. The JSSW provided its services until 1976. Thereafter its efforts have been sporadic. It followed its ELINT endeavours with its SJ spacecraft. Its first SJ launch in 1979 was a failure. Its SJ-2, second generation ELINT satellites provided services for a short while and thereafter followed a long hiatus, with the next SJ not being launched until 1994. Additionally, in September 1990, China launched a pair of DQ-1 (*Da Qi*/Atmospheric) minisatellites referred to as its third-generation satellites. These doublets were launched with FY (*Feng Yun*/Wind and Cloud) meteorological satellites. The mass and orbital parameters of these satellites were known to be consistent with ELINT applications. China's SZ spacecraft manufactured by the (SAST) Shanghai Academy of Space flight Technology under Project 921-1 for China's manned space program currently provides a substantial ELINT capability. In fact,[7] the latest mission involving the SZ-5 orbital module is believed to carry an optical reconnaissance camera with a ground resolution of 1.6 metres. The Chinese are also known to have attempted to enhance their SIGINT (Signals intelligence) and COMINT (Communications Intelligence) capabilities by trying to procure two APMT (Asia Pacific Mobile Telecomm) satellites in the open market from Hughes Space & Communications in California in 1996. These were meant to give the Chinese capability to eavesdrop electronically on telephone and mobile phone conversations not only in China, but also in up to twenty Asian countries ranging from Pakistan in the west to Japan in the north, and from northern China down south till Indonesia.

Interestingly the above deal raised American concerns following intelligence reports about General Shen Rongjun (a Chinese Army General who oversaw China's military space programmes) saying that he planned to emphasise the role of these satellites in information-gathering activities. Thereafter, in February 1999, the US Department of State formally denied Hughes an export licence for the APMT satellites and the contract was cancelled. Notwithstanding the above, the Chinese continue to seek similar telecomm satellites from European manufacturers and are likely to acquire geostationary satellites with secondary capabilities for intercepting not only mobile phone calls, but also other telecommunications using these services.[8]

Military Satellite Communications

With regard to military communications, since the PLA (People's Liberation Army) was allotted only limited channels amongst China's eleven communication satellites, it attempted to rectify the situation and proposed a network of defence communication satellites. Its FH-1 (*Feng Huo*-1) military communication satellite (first of the series) was launched in January 2000, which consists of the *Qu Dian* C[4]I (Command, Control, Communications, Computers and Intelligence) system. The network as per its registration with the ITU (International, Telecommunication Union) would consist of up to five satellites, China Sat 21-25. This network would enable PLA commanders to communicate with their in-theatre forces in near real time,[9] and also enable data transfer with all units under joint command in addition to providing the Chinese military with a high speed and real-time view of the battlefield, thereby enabling effective command and control. The Chinese military describes the new tactical information system component of the *Qu Dian* system as being analogous with the American JTIDS (Joint Tactical Information Distribution System). The satellites would reportedly provide the military with both 'C' and UHF band communications. Thus once fully deployed, the FH series constellation would establish space-based military tactical communication networks to support Chinese military operations.

Satellite Imagery

China's imagery capabilities are again a mirror reflection of its dual-use programmes. Its imagery programme focuses on both military reconnaissance and on Earth resources and disaster management. China became the third country, after the US and Russia, to launch an Imagery Intelligence (IMINT) satellite when its first FSW spacecraft was launched in 1975. A total of fourteen of these spacecraft were launched until 1994. An improved version, the FSW-2 was first launched in 1992, followed by another in 1994 and a third in 1996, and a more recent launch in 2003. However, these FSW satellites were known to physically eject film capsules for recovery on Earth. China's FY-1 series thereafter were an improvement on the FSW's capabilities, but usage of data links to relay electro-optical imagery back to Earth by radio signal began only with the ZY series of 2000.

During the interim period (1996–2000), China had surprisingly not launched an IMINT satellite. It is generally believed that China had met its limited IMINT requirements through commercial purchase of imagery from Russia. Since then the ZY system was in place to fulfill requirements typical to its immediate needs. The ZY series developed in conjunction with Brazil had made significant advances over the FSW series including data being downlinked rather than being film dropped. Its key sensors include

- CCD cameras with 19.5 metre resolution and 113-kilometre swathe.
- Wide-field Imager with 258-metre resolution and 890-kilometre swathe.
- Multi Spectral Infra-Red scanner with 78–156-metre resolution and 119.5–kilometre swathe width.

Thus with regard to satellite imagery, China has adequate capabilities and would definitely put its space forces to monitor defence developments in its area of interest. The Chinese also seem to be developing capabilities to counter overhead satellite reconnaissance capabilities. They have developed through practice a method with which a division combines hiding, lying low, and drilling, with the tactics of moving, deceiving, and harassing. These actions are sufficient to mask detection from both airborne and space-based surveillance.[10] It can thus be inferred that if the Chinese could task space assets to attempt detection of troops, they must also have trained its imagery analysts to examine claims and counter claims on satellite detection, satellite deception, counter camouflage, etc.

In the near future, the Chinese also plan to field a constellation of space-based reconnaissance systems with near real-time intelligence capabilities to support its military. This constellation shall comprise four Radar and four Optical reconnaissance satellites and is expected to be in place by 2010.

Micro-Satellites

China is conducting extensive studies and is seeking foreign assistance for small satellite constellations. It has already lofted a number of such satellites including the SJ-5 (Practice-5) in 1999, Hy-1 (Ocean-1) in 2002 for Ocean-surveillance, etc. Other missions for satellites of this class may include Earth observation, communications, navigation, etc. In the arena of micro satellites—small spacecraft that tip the scales at a little over 200 pounds (100 kilograms)—China is developing this class of spacecraft for remote sensing and networks of electro-optical and radar satellites. A joint venture between China's Tsinghua University and Britain's University of Surrey in building the *"Tsinghua"* system—a constellation of seven mini-sats with 164 feet remote-sensing payloads is also on the anvil. Russia launched the first satellite in June 2000. Later satellites in the series are expected to have greater resolution.[11]

China's demonstrated capabilities in micro and nanosatellites would provide it with a range of capabilities. Apart from ASAT operations, these would provide it with its much-needed 'launch on demand' capability by enabling surge capabilities during crises. Such microsats with their inherent versatility could fulfill multiple roles of reconnaissance, communications, ELINT, maritime surveillance, etc. by being deployed in swarms and constellations typical to requirements in designs like chevrons, triangles, squares and circles. These would provide in-depth redundant coverage of 'areas of interest'. Thus, these small cheap satellites, apart from enhancing the abilities of conventional forces and enabling space-support missions, would also bolster counter-space operations by their ability to double up as parasitic satellites. In view of the foregoing deliberation on Chinese eventual capabilities in space for intelligence, targeting, reconnaissance, etc. within Asia and the Pacific would amount to belabouring the obvious. To begin with, China's improved space-based surveillance positioning and targeting capabilities have already decisively altered the military balance of power within Asia, and a further launch of satellite constellations shall serve to accentuate the imbalance more perceptibly.

Satellite Navigation (SAT NAV) Systems

China's development of an independent satellite navigation and positioning system referred to as Twin Star of *Beidou,* was also in response to the declared defence requirements.[12] China's possession of such a system would help correct the historically low accuracy of its ballistic missiles. The use of terminal guidance systems and navigation satellites would enable positioning updates, allowing for midcourse guidance corrections. This would be especially helpful for ballistic missiles launched from platforms like the DF-31. A study conducted by the Rand Corporation estimated that by using GPS, the targeting accuracy of China's ballistic missiles could be improved by 20–25%.[13] Additionally, China has also invested up to $230 million in the European satellite navigation system of 'Galileo'. Access to Galileo's premium Public Regulated Service (PRS) which is intended for military applications and operates on a different frequency from America's GPS would confer immense strategic advantage and navigational redundancy in times of conflicts and crises.

Apart from missiles, China's aerospace industry has also sought to integrate satellite navigational systems into its fighter aircrafts and helicopters. China has also decided to equip all its new fighter aircrafts (both under development and procurement) with Sat Nav systems. These coupled with Chinese efforts to integrate imagery and Sat Nav systems into conventional forces would enhance its force capabilities tremendously.

Such enabling of targeting accuracy by Chinese satellites assumes greater significance in view of the fact that the Chinese nuclear doctrine emphasises a Chinese retaliatory strike against counter-value targets (enemy cities) rather than against counter-force targets (enemy missiles/forces). This is because counterforce targeting requires the use of highly accurate ballistic missiles, preferably with MIRV's (Multiple Independently targeted Re-entry Vehicle)—two technologies that China currently lacks in its operational ICBMs.[14] Without improved accuracy or MIRV, Beijing's force of only 20–30 ICBMs would lose its threat relevance subsequent to an effective deployment of the US BMD (Ballistic Missile Defence) umbrella. Further, if the umbrella were to accommodate China's Asian rivals like India and Japan, China's domineering military posture would be effectively reduced.

Finally, China's investing in multiple Sat Nav systems like the

American GPS, Russian GLONASS, European Galileo and its own indigenous BNTS would confer on it immense Sat Nav redundancy and strategic advantage. Details on Chinese satellites known to have been launched for military applications are given in Figure 6.2.

Chinese Satellites Launched for Military Applications

Type of Satellite	Common Name	Official Name	Launch Date	Remarks
Communication	Zhongxing-22 (Zhongxing means 'China-Star')	Chinasat22	25 Jan 2000	Also known as Feng Huo-1, China's dedicated military communication satellite with the Qu Dian C⁴I system on board.
	Zhongxing-20		14 Nov 2003	Dedicated military communication satellite.
Navigation	Beidou-01A	BNTS-1A	30 Oct 2000	Beidou means 'Big Dipper'.
	Beidou-01B	BNTS-1B	20 Dec 2000	
	Beidou-01C	BNTS-1C	14 May 2003	
ISR (Intelligence, Surveillance, & Reconnaissance)	Ziyuan-1	Ziyuan-2	01 Sep 2000	Full nomenclature– Zhangguo Ziyan (China Resource).
	Ziyuan-2	Ziyuan-2B	27 Oct 2002	
	Feng Yun-1C	–	06 Oct 1996	
	Feng Yun-2B	–	05 Oct 1999	
	Tsinghua	–	28 Jun 2000	
Signals Intelligence	Shenzhou-4	–	29 Dec 2002	
	Shenzhou-5			

Figure 6.2

Anti-Satellites (ASAT)

Chinese R&D on fundamental technologies applicable to an ASAT weapon systems has been ongoing since the 1960s. Under the 640 programme, the space and missile industry's Second Academy, traditionally responsible for SAM development set out to field a viable anti-missile system consisting of a kinetic kill vehicle, high powered laser, space early warning, and target discrimination system components. However, this programme was abandoned in 1980. The Chinese since then have concentrated on ASAT efforts in developing ground-based high energy weapons, ground- or air-launched interceptor missiles, parasitic satellites and 'hunter-killer' satellites that destroy their targets either through explosion or ballistic impact. Since 1998, there have been reports that China's Central Committee of Communist Party has been giving highest priority to the development of an anti-surveillance ASAT system. This system comprises ground-based lasers capable of damaging sensors of LEO (low earth orbiting) imaging satellites.[15] America's Cox report of 1999 also judged that China had the technical capabilities to develop CSS-2 into a direct ascent ASAT weapon. Some reports also mention the possible modification of China's solid fuelled missiles, the DF-21 or DF-31 as a direct ascent kinetic kill weapon. Presently China appears to be developing two kinds of attack satellites, conventional hunter-killer satellites and parasitic micro satellites. Reports indicate China's advancing proficiency in making advanced ASAT weapons like parasitic satellites.[16] These are micro-satellites designed to attach themselves to target satellites and capable of activation when required, to either jam or destroy the target satellite. It is claimed that these parasite satellites are capable of attacking satellites in low, medium or high orbit and are so small that they do not affect the target satellites normal functioning and hence go undetected.[17] Reports indicate such parasitic satellites are being developed by the Small Satellite Research Institute of the Chinese Academy of Space Technology.[18] Feasibility and maturisation of such programmes are likely to have significant implications for satellites the world over.

Prognosticating the Future

Prognosticating the future is fraught with difficulties on account of the immense pace of change in technological, geo-politic, economic and other developments. Thus while the long-term future would continue to be obscure in view of the prevailing developments, an assessment of the future advances and course most evident could be inferred and the same is undertaken. The future is inferred based upon known capabilities with the intention of attempting to assess the extent of Chinese space militarisation/weaponisation in the next decade and highlighting the emergent need for assessing the consequent implications and options. By 2015, the following space developments are most likely.

- A Chinese lunar base experiencing in Helium extraction and refinement techniques to augment its fossil fuel resources and other energy requirements.

- Operationalisation of China's 'Feng-Huo' military communication constellation.

- Operationalisation of its indigenous *'Qu-Dian'* C4I system for its military requirements.
- Operationalisation of its *'Beidou'* navigation, targeting and positioning system.
- Operationalisation of its IMINT constellation of 4 Radar and 4 optical satellites expected to be in place by 2010.

Broad Implications

Military Imbalance

It is common knowledge that numerical military force superiority is overwhelmingly in favour of China both within Asia and even in the world. Apart from the military force balance in both conventional and nuclear terms which is overwhelmingly in China's favour, the harnessing of military space capabilities would allow it to enhance its capabilities manifold. Additionally, China's military modernisation programme as inferred from its year 2004 white paper is overwhelmingly focused on enhancing its aerospace capabilities in terms of modern combat aircraft and space based force-multiplication systems, though it already possesses formidable capabilities.

China's military modernisation with its inevitable spin-off to Pakistan is drastically altering the balance of military capabilities, especially in the fields of air power and space-based capabilities. Apart from the aforementioned space-based capabilities, by 2010 it is estimated that China would dramatically enhance its capabilities in air power with acquisitions of over 350 Russian SU-30s and another 350 Chinese licence-produced SU-30s as opposed to India's 90 odd SU-30s and 45 Mirage-2000s by that time.[19] Even taking into account the premise that inventories of military air and spacecraft would be an imperfect metric of a nation's intent or capabilities since inventory by itself would need to be supported by concurrent doctrines, strategy, ground integrability, etc. the fact exists that the amount of literature spawned by Chinese defence analysts and the CAMS on the subject are reflective of Chinese emphasis on correct comprehension, operationalisation and integration of space capabilities into terrestrial war-fighting capabilities. Chinese emphasis on military cost-cutting by reducing manpower and harnessing modern technology are reflective of their level of comprehension of the capabilities afforded by space.

Proliferation

China has a known record of proliferating missile technology to other countries, especially North Korea and Pakistan. The Chinese space programme facilitates this effort both internally, by pooling China's civilian and military know-how, and externally, by serving as a political cover for foreign technology transfers.[20] This is of relevance, especially when considering Pakistan's and North Korea's unique position as strategic allies of China. Chinese military technology transfer to Pakistan and North Korea in addition to its history of promoting Pakistani military capabilities to keep India engaged on the one hand and advancing North Korean ambitions on the other to keep Japan and the US on tenterhooks are clear indicators of the fact that China shall be a willing patron in proliferating or sharing space resources and technology to Pakistan and North Korea, especially during crises and conflicts. It can thus be foreseen that their space collaboration shall follow on the same lines as in the case of nuclear and missile proliferation. Table 6.1 on recent missile flight tests would also serve to further validate the Chinese–Pakistan–N. Korea collusion nexus.

Recent Missile Flight Tests

Msl flt test (2003–04)	Pak Version	Original Chinese/N. Korean nomenclature	Range (km)	Payload (kg)
03 Oct 03	Hatf-3	M-11/DF-11/CSS-7	280	500
14 Oct 03	Hatf-4	M-9/DF-15/CSS-6	750	500
29 May 04	Hatf-5	No-Dong	1,300	1,000
04 Jan 04	Hatf-5	No-Dong	1,300	1,000
09 Mar 04	Hatf-6	M-18	2000–2500	1,000

Table 6.1

Economic Rivalry

Space has become valuable territory, both commercially and strategically. It is a universally accepted norm that human self interest and by extension, state and national self-interest are the overriding factors driving wealth, economies and by extension conflicts of interest. Analogous to the theories of "commons" pertaining to mutual economic exploitation of the seas by nations in the early eighteenth and nineteenth centuries which soon led to disagreements and conflict and later control of the seas when individual nations sought to seek advantages for themselves and other nations sought to challenge and wrest these advantages, space is also transiting through such a phase and it would not be long before nations seek to wrest equal (or unequal) advantage by means fair or foul. Both China and India have full-scale space programmes aimed at economic enhancement of their respective nations. Both have their respective advantages of cheap labour, info-tech resources, support infrastructures, and vibrant competitive economies thereby allowing them great potential to compete for domination of the lucrative worldwide space market. For example, the launch costs of the Chinese were the lowest in the commercial market until the Indians came out with cheaper alternatives displacing the Chinese monopoly as the most affordable launch providers. Thus, while competition for the lucrative space market between India and China has already begun, the same would not be indicative of the inevitability of large-scale military conflict in space. Nevertheless, the eventuality of disagreements, conflicts of interest, etc. snowballing into major endeavours attempted at denying space advantages to the opposing rival cannot be totally ruled out. In fact, small-scale conflicts of interest are already under way within Asia. Present areas of conflict in space deal with affairs such as geostationary spacing or spectrum allocations, interference in communication signals, denying of TV broadcasting, jamming of GPS during conflicts, etc.[21]

Energy

Asian economies are greatly dependent on energy for perenially sustaining their growth and the continent, apart from being home to the world's energy "demand heartland" comprising Japan, China, India, etc. is also the energy "supply heartland" of the world with its energy resource periphery extending from central Asia, Persian Gulf, North Indian Ocean, South China sea, East China Sea, etc. Energy demands in future are expected to increase exponentially as opposed to shrinking supplies and the consequent implications of the same on the Asian security calculus is known to be profound. Helium 3 deposits on the Moon are now touted as the next credible alternative to Earth's depleting fossil fuel reserves. The Moon is known to contain 10 times more energy in the form of Helium 3 than all fossil fuels on earth and while present-day reactor technology for conversion of Helium 3 into energy is in its infancy, technology and economic dynamics would soon provide enough impetus for "cost-effective" harnessing of these energy resources. Considering

China's declared territorial ambitions with regard to the moon and Helium extraction and its consequent implications on the other Asian space-faring nations of India and Japan (who are equally energy starved) as well as the US, the stage is already set for migration of terrestrial conflicts into space.

Territorial Claims in Space

Apart from seizing control of the ultimate "high ground" during conflict situations, Chinese space scientists in 2002 were also reported to have urged their government to accelerate acceptance of a proposal to develop infrastructure in space, stake territorial claims on outer space, and develop the "space territory" as part of national strategy. Notwithstanding the treaty obligations of the UN 1967 OST (Outer Space Treaty) which declares outer space as the "province of all mankind", Chinese space specialists are known to have argued that by virtue of having "vehicles that take up positions in space and the ability to possess part of the space resources", a country would effectively extend its three territorial claims—land, sea and air—into space; thus the claim of the "fourth territory". Chinese space scientists also recognise that there is fierce competition for space resources, but most nations do not currently have the capability to be participants. The implications of Chinese territorial ambitions in space are ominous, especially considering China's plans to build and develop lunar facilities. Chinese plans for explorations of Helium 3 on the Moon, which could fulfill its energy needs and its consequent implications for economic advancement could be the driving factors for such territorial ambitions in space.[27]

Conclusion

Thus while China has relatively moderate military space capabilities compared to the US and Russia, it is clearly the undisputed military space power in Asia. In pursuit of its 'force-enhancement' or 'information-supporting' missions, it has developed satellites for communications, surveillance and reconnaissance, navigation and positioning, etc. which have further enhanced and augmented its formidable military capabilities. Its reconnaissance capabilities in particular permit it to monitor neighbourly defences and form near-accurate ELINT and SIGINT 'OrBats' (Order of Battles). Its navigation satellites are set to ensure and enhance the accuracy of its ballistic missile capabilities. It is also suspected to be making substantial progress in its 'space-control' or 'battle-field combating' roles by actively pursuing ASAT capabilities. This is surely cause for concern especially considering the fact that the Asian land mass includes up to four of the eight nuclear weapon states, most of the world's fastest-growing economies, most of the world energy resources, consequent energy rivalries, and numerous other factors nurturing conflict and tensions rather than mollifying them. In such a volatile situation, a regional imbalance of power would only serve to exacerbate regional tensions, suspicions and rivalries and most likely trigger a space weaponisation race in addition to the prevailing nuclear competitions and tensions.

PAKISTAN'S MILITARY SPACE PROGRAMME
Origin and Development

More than four decades ago in 1961, Pakistan set up the Space and Upper Atmosphere Research Commission (SUPARCO) with the announced goal, not yet attained, of launching Pakistani satellites aboard Pakistani rockets. In June 1962, the United States launched the first rocket from Pakistani soil. The launch used a combination of two US rocket motors, the Nike and the Cajun. Fired from Sonmiani Beach, 50 km west of Karachi, the rocket reached an altitude of almost 130 km. The US space agency NASA hailed the launch as the beginning of "a program of continuing cooperation in space research of mutual interest". The NASA–SUPARCO cooperation agreement called for the training of Pakistani scientists and technicians at NASA space science centres. Before the June 1962 launch, NASA had begun to train Pakistani scientists at Wallops Islands and the Goddard Space Flight Centre. NASA also set up fellowships and research associate programmes at American universities for "advanced training and experience".[23]

Europe also aided Pakistan's early rocket development. France transferred technology to manufacture sounding rockets and German firms assisted in space research and supplied several tons of ammonium perchlorate, an ingredient of solid rocket fuel. Great Britain also helped with sounding rocket launches. In fact, several SUPARCO personnel completed their master's degrees in engineering at England's University of Surrey—the institution that built and operated numerous small satellites such as the UO-9, UO-11 and UO-22 hamsats. While at Surrey, SUPARCO personnel worked on UOSAT projects with support from the Pakistan Amateur Radio Society, engineers who were hams at SUPARCO in Lahore began building a small hamsat during the last half of 1986. They used knowledge they'd gained at the University of Surrey to build their own satellite, and christened their new satellite *Badr*, after the Urdu word for "new moon".[24] However, the satellite was mocked by the international media, and some British papers claimed it to be the ditto of a British student's thesis in UK.[25] While Pakistan's first satellite was launched nearly three decades later in July 1990, its next satellite followed after another decade-long gap on 10 December 2001. The next urgent need to place a satellite in orbit was felt when the International Telecommunication Union (ITU) served Pakistan a notice regarding cancellation of its last space slot in the geo-stationary orbit. Five space slots were allotted to Pakistan by the ITU way back in 1984, but since Pakistan had failed to launch any satellite till 1995, it lost four of its allotted slots. In 1995, Pakistan again applied for and received five slots, however, the Pakistanis again failed to put a satellite in orbit, losing four of its slots in the process. To secure its slot, Pakistan finally leased a third-hand Hughes Global Systems Satellite (HGS3) for five years with an initial cost of around $4.5 million and named the satellite which originally began life as Indonesia's Palapa C1-"PAKSAT-1".

Space Organisation and Infrastructure

Pakistan's Space and Upper Atmosphere Research Commission (SUPARCO) was established in 1961, and started functioning in 1964. It implements space policy established by the Space Research Council (SRC), whose president is the Prime Minister. SUPARCO's programmes include the development and launch of sounding rockets, and satellite applications in the field of remote sensing and communications. SUPARCO is headquartered at the Arabian Sea port of Karachi and southern Pakistan, with additional facilities at the University of the Punjab at Lahore. Pakistan has undertaken a number of steps for consolidating and focusing its space programme in response to national priorities. A satellite ground station for reception of NOAA, Landsat and SPOT data was upgraded in the late 1990s. A national GIS Committee was constituted to bring about GIS standardisation. A Space Applications Research Centre has been commissioned at Lahore, and the Aerospace Institute, under SUPARCO, imparts training and educates space application experts. The SUPARCO Complex near Rawat is a comprehensive facility that tracks, acquires, archives, processes and analyses imagery from earth observation satellites such as Landsat, SPOT and NOAA. The acquisition zone of this facility extends to approximately 2,500 km around Islamabad, and its products are being provided to more than 70 public and private sector user agencies in Pakistan. SUPARCO's remote-sensing endeavour, RESACENT (Remote Sensing Applications Centre), Karachi is known to actively pursue a multi disciplinary remote-sensing application programme since 1973. Research studies based mainly on Landsat, SPOT data supplemented by aerial and conventional data have been carried out at the centre.[26] However, the Pakistani media believes that the performance of Pakistan's space programme and particularly that of SUPARCO has always been much below the desired levels.[27]

Launch Endeavours

Ever since the launch of the first Rehbar (Nike-Cajun) rocket on 7 June 1962, SUPARCO claims to have launched more than 200 rockets in the altitude range of 20–550 km. The Rehbar flights were basically aimed at acquiring meteorological data pertaining to wind velocity and direction at altitudes of 50–80 miles. SUPARCO's meteorology sounding rockets programme continued in the 1970s with Skua rockets being fired in March/April 1973 to conduct experiments of stratospheric winds and temperature measurements. In 1981, the head of SUPARCO announced plans to test a launcher in 1986, and by the mid-1980s, Pakistan had "established its own rocket production plant where rockets required for high-altitude scientific research are manufactured", according to then chairman of SUPARCO, Salim Mehmud. SUPARCO also built rocket-test facilities, chemical and propellant laboratories, high-speed tracking radar and a laboratory to work on telemetry. Despite SUPARCO's civilian orientation, the agency is believed to be involved in the development of short- and medium-range solid-fuelled ballistic missiles for Pakistan's military. SUPARCO has set up facilities for the complete manufacture of rockets and by February 1989, in an address at the National Defence College in Rawalpindi, Pakistan's Army Chief of Staff General Mirza Aslam Beg announced that two indigenously manufactured surface-to-surface missiles had been tested ('Hatf-I', a short range solid-fuelled missile with a range of 70–100 km, a payload of 500 kg and 'Hatf-II', with a similar payload but a designed range of 300 km). However, some analysts believe that the agency most likely used technology imported for the sounding rocket programme in the development of the Hatf-I and Hatf-II short-range ballistic missile programmes in the late 1980s and early 1990s. In fact, analysts on the subject aver that Pakistan's Hatf is based on France's sounding rocket technology.[28]

In July 1991, the United States sanctioned SUPARCO and two other Chinese entities for missile-related proliferation activities. In 1995, the United States warned that SUPARCO was seeking equipment from various European suppliers for Pakistan's ballistic missile programme. Among other items, SUPARCO has attempted to procure composites, specialty alloys, and production and testing equipment for rockets, including electronic beam welding equipment which is used for specialised welding on a missile airframe. In March 1996, Taiwanese custom authorities intercepted 200 tons of Ammonium Perchlorate (AP) bound for SUPARCO on a North Korean freighter. AP is an oxidizer used in solid propellants; the shipment originated in Xian in China. Hong Kong authorities also seized 10 tons of AP bound for SUPARCO in April 1996; the shipment originated in the North Korean port of Nampo and was shipped through Xian in China. Another 10-ton shipment of AP was intercepted by Hong Kong customs in December 1996; the shipment apparently originated in North Korea and was routed through China.[29]

In early July 1997, Pakistan's National Defense Complex (NDC) tested a Hatf-III/Ghaznavi/M-11 ballistic missile from SUPARCO's flight-test range at Sonmiani beach, which is located 50 km north-west of the port city of Karachi. Subsequently, the NDC also tested the Hatf-IV/Shaheen-I (possibly M-9) ballistic missile from Sonmiani in April 1999.

Thus SUPARCO has always been involved in the development of solid-fuelled short and medium range ballistic missiles with Chinese and North Korean assistance. An examination of missile flight tests in the recent past (2003–04) would also be reflective of the growing Chinese–N. Korean–Pakistani nexus on missile proliferation.

Four decades since its inception, Pakistan is still in the process of developing its own SLV. However, in keeping with its penchant for bombastic pronouncements not matched by ground realities or achievements, Dr. Abdul Qadeer Khan, the father of Pakistan's nuclear programme, announced in March 2001 that Pakistani scientists were in the process of building the country's first SLV and that the project had been assigned to Pakistan's national space agency, (SUPARCO), which also built the Badr satellites. According to Dr. Abdul Majid, chairman of SUPARCO, Pakistan envisaged a low-cost SLV in order to launch lightweight satellites into low-Earth orbits. Dr. Khan also cited the fact that India had made rapid strides in the fields of SLV and satellite manufacture as another motivation for developing an indigenous launch capability.[30] According to an Islamabad news source, the SLV would be derived from an already available missile launching system, which may be an indication that technologies acquired for the ballistic missile programme would eventually be used to develop an SLV.[31] As of now, Pakistan's indigenous SLV programmes continue with no indigenous SLVs in sight in the near future.

Pakistani aspirations with regard to developing SLVs and corresponding ballistic missiles are ostensibly driven by its desire for a separate missile command structure premised primarily on land-based missiles on mobile launchers which would be the mainstay of Pakistan's nuclear force till such time a second strike capability and solid fuelled missiles are ensured.[32] This reflects strong doctrinal overtones of the Chinese strategy of compensating accuracy with mass since the present Pakistani nuclear doctrine is aimed at counter-value targeting (enemy cities) which requires less accuracy than counter-force targeting (enemy forces) with a 'one-rung escalation ladder' in terms of all-out war. In fact, it publicly proclaims that its counter-value targets are Indian urban and industrial centres, the critical ones being those within range of Pakistan's prevailing delivery systems.[33]

Pakistan's space launch facilities are located in southern Pakistan at Sonmiani Beach on the Arabian Sea. However, SUPARCO has only launched sounding rockets from this site, referred to as the Flight Test Range (FTR). These facilities are shared with the Pakistan Atomic Energy Commission's National Defence Complex, which uses the site to flight-test solid-fuelled ballistic missiles. SUPARCO also has a ground station near Islamabad and telemetry, tracking, and control stations located at Sonmiani Beach, Karachi, and Lahore.

Satellite Programmes

Badr

In a bid to manufacture their first indigenous satellite, SUPARCO engineers with support from the Pakistan Amateur Radio Society began building a small hamsat during the latter half of 1986. They used knowledge gained at UK's University of Surrey to build their first 150-pound satellite, the *Badr-1/Badr-A*. *Badr-1* was to have been ferried into space aboard a US space shuttle, but that plan changed after the 1986 Challenger explosion delayed further US shuttle flights. China subsequently agreed to launch *Badr-1* on one of its Long March rockets. In 1989, Pakistan registered the planned satellite with the International Frequency Registration Bureau. The spacecraft was shipped to China's Xinxiang Launch Centre in 1990. The tiny spacecraft was launched as a secondary payload on a Chinese Long March 2E rocket from the Xinxiang Space Launch Centre on 16 July 1990. The primary payload was Australia's Aussat satellite. Originally designed for a circular orbit at 250–300 miles altitude, *Badr-1* was actually inserted by a Long March rocket into an elliptical orbit of 127–615 miles. One of eight hamsats sent aloft in 1990 around the world, *Badr-1* circled the globe every 96 minutes, passing over Pakistan for 15 minutes three to four times a day.[34] *Badr-1* offered one radio channel for digital store-and-forward communications, its transponder uplink was near 435 MHz, and the downlink was near 145 MHz. However, contact with the spacecraft was lost on 20 August 1990. *Badr-1's* orbit was so low that it could not sustain itself in space for more than 146 days and it burned up in Earth's atmosphere on 9 December 1990.

Eleven years after *Badr-1,* Pakistan's second satellite, *Badr-2* was finally launched on 10 December 2001. It was carried to space by a Zenit-2 rocket from Russia's Baikonur Cosmodrome in Kazakhstan. Satellites from other countries that flew alongside *Badr-2* on the Zenit booster were Meteor-3M 1, Kompass, Maroc-Tubsat and Reflector. Initially, SUPARCO planned to launch the second *Badr-2* satellite during 1993. However, the target could not be achieved. A plan to launch *Badr-2* in 1994 also did not materialise, and it was hoped that it would ride in 1995 or 1996. The launching was then slated to be done from the Baikonur Cosmodrome, Kazakhstan in late August 1999 on a Zenit-2 rocket [the main satellite to be launched was Russian]. Four satellites—

one each from Pakistan, Malaysia, Morocco and the US—were mounted on the bigger Russian satellite. As per convention the satellite would be known as *Badr-2* after its launch. The anticipated launch date subsequently slipped to early 2000 and then to 2001.[35]

Badr-2 is a gravity gradient stabilised small Earth Observation satellite designed by Space Innovations Limited [SIL] of the United Kingdom. While spacecraft sub-systems are SIL designed and manufactured, the spacecraft integration was undertaken by SUPARCO of Pakistan, demonstrating the use of relatively inexpensive microsatellite missions in the field of space technology transfer. Most of the equipment used in the satellite was acquired in Pakistan to stimulate the local software industry. *Badr-2* was known to carry the following four experimental payloads:

- Earth Imaging CCD camera
- Battery End-of-Charge detector
- Radiation Dosimeter
- Store-and-forward communications.

In view of its payloads, there are four objectives of the project. First, *Badr-2* has a CCD camera through which the earth's imaging can be done. Second, there is equipment in the satellite with the help of which signals sent to it, that is, e-mail, etc. could be stored. These signals might be forwarded later for onward delivery. The satellite would also be able to measure the radiation through a dosimeter. The last objective is to carry out the battery-end-of-charge-detection. The successful operation of the CCD camera on board the *Badr-2* satellite would be a first step towards the acquisition of know-how for taking pictures of Earth from specialised digital cameras.

Remote-Sensing (Spy Satellite) Endeavours

While *Badr-2* is a modest attempt at Earth observation, Pakistan's fledgeling space programme has long been oriented toward remote-sensing applications. A data processing infrastructure has been established to exploit Earth observation data transmitted by Landsat, NOAA, and SPOT satellites. In fact, the Russian daily *Vermya Novostyei* had revealed that Pakistan has offered US$130 million to Russia to develop and launch a spy satellite to keep an eye on India and other neighbouring countries. It said Pakistan was exploring the possibility of placing orders with the

Russian space industry for developing and launching a remote-sensing satellite with high resolution cameras. Ousted Prime Minister Nawaz Sharif signed an $800,000 deal to launch a mini-satellite with Russia in the 1990s, but the deal never materialised.[36]

In April 1998, Pakistan, China, Iran, South Korea, Mongolia, and Thailand signed a "Memorandum of Understanding on Cooperation in Small Multi-Mission Satellite and Related Activities". The small multi-mission spacecraft (SMMS) project aims to develop a satellite for civilian remote-sensing and communications experiments. The SMMS satellite is scheduled for launch into sun-synchronous polar orbit by 2005. It will carry a hyper spectral imager and two wide-swatch charge-coupled device cameras, developed with the help of Iran, as well as an experimental telecommunications system. The sharing of space-imaging technology made possible by the MoU could help give Pakistan an autonomous military reconnaissance capability. Pakistan is also known to be working with China on a regional remote-sensing initiative.[37]

Additionally, in November 1999 Dr Abdul Majid, Chairman Pakistan Space and Upper Atmosphere Research Commission (SUPARCO), said that Pakistan was planning to develop its own indigenous 'earth observation satellite' within a period of two to three years, although he did not offer any further details.[38] By April 2002, SUPARCO had announced that it planned to launch its high-resolution Earth Observation Satellites System (EOSS), in three to four years using a three-phased approach,[39] and by January 2003, Pakistani President Pervez Musharraf had also declared that Pakistan must also have an observation satellite. Pakistani literature on the subject is also of the opinion that "Pakistan cannot afford to rely on western satellites for intelligence gathering during war and peace. If Pakistan can develop its nuclear capability and the related delivery systems (ballistic missiles) on its own with some foreign help, then indigenous development of space programme too can become a reality. This will require resource allocation to achieve the desired goal".[40] From the foregoing it is apparent that Pakistan in the near future would be making all-out efforts to place some kind of imaging capability in place.

PAKSAT-1

PAKSAT-1 began life as a series 601 satellite built by Boeing. It was launched on 31 January 1996 from Kourou in French Guiana by an Atlas

2AS booster for Indonesia under the nomenclature *Palapa C1*. The 601 series, however, were not very well-known for reliability in orbit and several had failed with electrical related problems. "At least three seem to have been total losses, and more are damaged..." as per columnist Muhammad Irshad of Pakistan's defence journal. On 24 November 1998, less than three years after launch, *Palapa C1* experienced electrical problems. This problem was different from those afflicting some of the other satellites, in this case the battery charge controller failed. This left the satellite with no way to recharge its onboard batteries during an eclipse period. This is a major fault on a satellite because no backup power is available during the eclipse period which happens twice a year. On undamaged satellites, the batteries provide the power needed to run the satellite for the few days every six months when the sun is in the wrong place to illuminate the solar cells. Due to onboard failure, the transponders need to be switched off during this time.

As a result, *Palapa C1's* mission to provide telecommunication links to the Indonesian islands remained incomplete and the satellite finished its utility after less than three years. An insurance claim was settled, and the satellite ownership passed to the insurance company. In January 1999, Hughes Global Services purchased the satellite from the insurers, renaming it HGS3. Kalitel, a US-based company, leased HGS3 from Hughes, and in December 2000, it was rechristened *Anatolia-1* and moved to Turkey's registered slot at 50 E. It reentered service on 12 February 2001 as Turkey's *Anatolia-1* providing discount KU and C band coverage of Europe, Africa and Asia. By the beginning of August 2002, it was announced that the Turkish lease had ended, and the satellite was leased to Pakistan for five years.[41] However, Pakistan's science ministry admits that problems with the battery pack persist, which do not allow the batteries to provide energy to the payload during the eclipse period of 88 days a year, nevertheless, the availability of the satellite transponders is claimed at more than 96% despite outages during the eclipse period.

On 3 July 2002, announcing a federal cabinet decision, Pakistan's Minister for Science and Technology, Dr Atta-ur-Rehman said that the country was acquiring the above-mentioned HGS3 on lease for five years at an initial cost of $4.5 million. As borne out earlier, the deal was inspired by Pakistan's concern about permanently losing its fifth and last 38 degree east slot since it had already lost four of the five slots originally allocated to it in 1984 by the ITU, which regulates satellite-related matters world wide. Indeed, Atta-ur-Rehman said it was because of the urgency of securing its last slot in space that Pakistan opted to get a "used but cheapest available satellite" on lease. While Rehman stressed that the decision was motivated solely by commercial concerns, experts noted that having a satellite would also enable Pakistan to beef up its defence communications at the very least. Air Vice Marshall Azhar Maud, chairman of the National Telecommunications Corporation (NTC), himself said that a geo stationary satellite could be used to secure defence communication, act as a lookout for a missile attack and detect any nuclear detonation or explosion. M Nasim Shah, secretary of SUPARCO also said that the technology is vital for making the nuclear command and control mechanisms "credible".[42] After *PAKSAT-1* successfully commenced operations in January 2003, President Pervez Musharraf declared that Pakistan must also have an observation satellite. Pakistan announced plans in late January 2003 to launch an indigenous satellite to replace *PAKSAT-1* within three years, and President Musharraf stated that this satellite would meet communication and earth observation needs for Pakistan's national interests.[43]

From the foregoing, it is obvious that Pakistan's *PAKSAT-1* is an intermediate arrangement till it gets a viable military or dual-use system in place for communications and earth observation. This is validated by the fact that Karachi-based SUPARCO has issued notices to potential bidders to provide a business plan and technical help in evaluating a future Paksat-1R satellite, claiming that the new spacecraft would be launched by December 2007.[44]

Communication and Broadcasting Programmes

Like India, Pakistan's satellite communication programme goes back to the early 1980s when SUPARCO, in collaboration with Hughes, conducted a feasibility study defining the broad parameters of what was to be called PAKSAT. The project's estimated cost was 400 million dollars. PAKSAT, however, remained on the drawing boards, that is, until officials realised that Pakistan was about to be shut out of the GSO system.

Pakistan also feels that in an emerging broadcast environment, it desperately needs a communication satellite of its own. Pakistan's government-controlled TV channel PTV itself realised the importance of satellite when in 1992, it rented a transponder on the AsiaSat 1 and started beaming programmes to India and other parts of South Asia and Middle East.

To reach Europe and North Africa, it later rented a transponder at ThaiSat. Such efforts, however, were no match to those of India, which by then was beaming a large number of its channels across the region and elsewhere. And as more than half a million dish antennas mushroomed atop houses, dozens of Indian channels and programmes were entertaining more and more Pakistanis. Some Pakistanis viewed this as an Indian cultural onslaught, and by the late 1990s that Islamabad began making moves to keep out the so-called "Indian invasion".

The government started by regulating the cable companies. Then, in December, as ties sank to a new low after an attack on the Indian Parliament, Islamabad banned the transmission of Indian television channels into Pakistan.

ISNET (Inter Islamic Network on Space Sciences and Technology)

Pakistan's obsession with leadership of the Islamic world (as in the case of its "Islamic bomb") is reflected in its establishment of ISNET in 1986, with member countries including Pakistan, Malaysia, Indonesia, Jordan, Syria, Bangladesh, Bahrain, Brunei, Kuwait, Senegal and Cameroon. ISNET is headquartered at SUPARCO headquarters, Karachi, always headed by the Chairman SUPARCO as its President, and Executive Director SUPARCO as its Executive Director, all-in-all a Pakistani endeavour aimed at "promoting the advancement of Space sciences and Technology in the countries of the Islamic World". As in the case of the Muslim Ummah's first "Islamic bomb", it endeavours to position itself as the Muslim Ummah's leader in space sciences and technology, incognizant of the fact that in the "Islamic World" countries like Indonesia, Malaysia, etc. have early well-developed space programmes, have a larger number of satellites in orbit, and also possess the economic means to buy commercial space technology based on their requirements rather than scout around for the cheapest stuff as in case of Pakistan.

Doctrinal Drivers

Apart from Chinese hardware proliferation, the impact of Chinese doctrinal thought influencing Pakistani mind-sets in developing space capabilities is profound considering the fact that apart from China,[45] Pakistan is the only nation in the world which thinks in terms of carving out national territory in outer space notwithstanding the Outer Space Treaty of 1967 which considers space as the province of all mankind. For example, the Pakistan Air Force (PAF) Chief, Air Chief Marshal Mushaf Ali Mir has gone on record stating that "it is conceived that as national frontiers on ground, there will be national frontiers in space that nations will need to defend, as indeed they defend their national air space".[46] Secondly, the Chinese emphasis on space-enabled "Informationalisation" as well as Electronic Warfare (EW) is reflected in the statement of SUPARCO's secretary M. Nasim Shah that though Pakistan's space programme is presently not in a position to engage in EW, space technology could help towards that end and that SUPARCO was embarking upon a programme of communication satellite development which would allow it in developing the capability for EW. Thirdly, like the Chinese, Pakistan is overtly an ardent public advocate of the peaceful uses of outer space while covertly it attempts to shop for and obtain space-based capabilities like spy-satellites and delivery vehicles. In fact, SUPARCO's secretary contradicts himself by stating on the one hand that Pakistan is an ardent advocate of the peaceful uses of outer space...and on the other hand replying to a query on the SUPARCO's need for observing the UN conventions and avoiding militarisation of space, the secretary responded that such conventions were no longer binding because of America's decision on going ahead with BMD systems and its general intransigence on the subject. Lastly, the secretary's pronouncements in response to queries regarding the space-based options available to Pakistan following Indian troops concentrations on its borders and unreliability of Western (particularly American) satellite systems as well as low performance of SUPARCO was that they had their own intelligence systems for facilitating counter-value targeting.[47] This could be construed as being indicative of either Chinese space collusion or Pakistani disinformation campaigns since the only other Asian space-faring nations like Israel or Japan were unlikely to provide support against India.

Conclusion

In view of the foregoing it can safely be surmised that while Pakistan has immense ambition (both military and civil) with regard to space, it currently lacks the wherewithal for the same. However, in view of its pronouncements and ambitions, once the riders of adequate finances and technology are overcome, Pakistan may go in a big way towards fulfillment of its cherished goals with respect to space-enabled capabilities. Pakistani ambition coupled with Chinese and North Korean collusion may in the near future allow at least a partial consummation of its desired space capabilities.

JAPAN'S MILITARY SPACE PROGRAMME

Resilience is an outstanding characteristic of the Japanese, profoundly manifest in their space programme which has experienced sporadic bouts of spectacular failures bringing it almost to the brink of collapse only to bounce back stronger and more resilient than ever before. This attribute of the Japanese people is reflected in its space programme, begun post-World War II by a people reeling under the degradation and humiliation of nuclear bombardment and military defeat. For a programme begun under such trying circumstances, the Japanese space programme starting with a 200-grams pencil rocket has come a long way in the past five decades and the Japanese are recognised as a formidable space power to be reckoned with not just in Asia but in the entire world.

The Genesis

Following Japan's defeat in World War II, the General Headquarters of the Allied Powers had banned Japanese armaments completely. It is universally known that the Japanese Constitution drawn post-World War II under the auspices of General Douglas McArthur's victorious American forces places strong restraint on Japan acquiring extensive military capabilities. Additionally, a resolution adopted by the Diet on 9 May 1969 in the House of Representatives relating to the basic principles of development and use of space, approved a narrow definition of Japan's space development policy which prohibits the use of advanced space technology by the Japanese Defence Agency (JDA), thereby committing Japan to solely peaceful uses of outer space. It explicitly states that "the development and utilisation of objects that will be launched into outer space and rockets that will be used to launch them shall be limited to peaceful purposes".[48] However, the 1969 space resolution was altered by the then Prime Minister (PM) Yasuhiro Nakasone to make it possible for Japan to use some space technologies for military purposes provided that the technology was commercially available. Hence Japan could use its JCSAT and Super bird satellites, both run by private companies for the land and sea forces to communicate with each other.[49]

The above, however, was perceived to be inadequate by the Japanese in view of North Korean belligerence as well as Chinese military advances, hence by 1994 Japan began a serious reconsideration of its long-held policy prohibiting the use of space for military purposes. The JDA, SAC and the non-governmental "Defense Research Center" all issued findings that non-lethal, particularly photographic reconnaissance, military space missions were a logical extension of Japan's space and national defence activities.

In keeping with the spirit of these deliberations and findings, in 1996 a joint commission of the Liberal Democratic Party (LDP) research committee on foreign affairs and security met with members of the NEC Corporation, which manufactured 41 of the 63 Japanese satellites, to discuss the feasibility and costs of a satellite reconnaissance system. Press reports were also indicative of the fact that the newly formed Japanese Intelligence Agency was lobbying for "spy-satellites", as was the foreign ministry in order to lessen dependence on the US. The Ministry of Foreign Affairs (MFA) was also known to have employed a US consultancy to determine how other countries used reconnaissance satellites, and on the basis of this study, recommend a system for Japan.[50]

The NEC study suggested a $2.4 billion price tag and annual operating costs of $200 million for a Japanese system. The following year (August 1997) the Science and Technology Agency (STA) presented a budget request to the MFA. Following negotiations between the two ministries, cabinet approval was given in mid-January 1998. Despite complaints from the US, the budget allotted $38,000 (¥5 million) to start development of a Japanese intelligence system.[51]

Prior to the launch of Japan's indigenous spy-satellites, Japan was known to rely on American and to a lesser extent European satellites for reconnaissance. For example, it obtained from the US (Land Sat) and

French (SPOT) satellite imageries with resolution of 30 and 10 metres respectively which were not militarily significant. The JDA to skirt both letter of the law and international sensitivities, had signed agreements to purchase imagery from commercial firms. The effort, dubbed the Geospace Information System had Tokyo-based companies tying up with US partners to buy sub-one metre resolution images as early as 1998. For example, Hitachi reportedly had signed a deal to resell imagery data from the US-based Earth Watch Inc and Mitsubishi Corporation had teamed up·with Lockheed Martin and Eastman Kodak to form Space Imaging Co. to sell similar products. While the MFA favoured Mitsubishi, the JDA had left the option open and was ready to purchase from either companies depending on the quality of the product. It was also speculated in some quarters that the JDA had been pursuing spy satellite technology under the guise of NASDA's ALOS which was ostensibly meant for mapping and environmental research.[52]

The importance of an integrated intelligence gathering satellite system was profoundly appreciated by the JDA and a domestic manifestation of its efforts was the JDA's Defence Intelligence Headquarters (DIH) in Ichigaya which included a new space imagery division and a staff of 1600 devoted to surveillance, planning, administration and threat assessment roles. The DIH was set up in 1997 under the Joint Staff Council to coordinate intelligence-gathering and analysis by the ground, maritime and air self defence forces. Additionally, keeping in mind the importance of satellite imagery analysis and its impact on the decision-making process, the Cabinet Satellite Information Centre (CSIC) was set up for the Information Gathering Satellite (IGS) project and images captured by the satellites were to be analysed by CSIC staff who started preparing for their duties up to five years ago and included about 200 government officials dispatched from various ministries and agencies, the defence agency, as well as around 100 others from the private sector.

Finally, on 31 August 1998, N. Korea ignited the simmering tensions, fears and concerns of Japan by launching a "Taepo-Dong" missile across Northern Japan. The first part of the missile fell into the Japan Sea and the second part (and probably third part) flew across the Japanese territory of Honshu and fell into the Pacific Ocean. It was then believed that the Japanese did not even know about the object's cross-flight until the US later informed them. However, recent press reports indicate that with its spy satellites and U-2 reconnaissance flights, the Americans did give the JDA plenty of warning before North Korea launched the missile. The JDA was also given time to divert an Aegis destroyer to the Sea of Japan so that its sophisticated radar could be used to track the missile's flight path.[53]

The North Koreans however contended that the object was a 'singing satellite' meant to broadcast patriotic songs. The US state department, subsequent to its own investigation had concluded that the suspected missile was a small satellite that had failed to reach orbit. But, the Japanese government was conclusive in its judgment of the object as a ballistic missile[54] and promptly stopped contributions to the Korean Energy Development Organisation (KEDO); the Japanese foreign minister released a statement with the secretary of state of the US and the South Korean foreign minister condemning the launch.

On 7 September, a week after the launch, the LDP started to discuss the possibility of introduction of multi-use satellites for information gathering. On the same day, the Shinto Heiwa and the Democratic Party also met to discuss the problem. By the next day, on 8 September, the security council of the LDP decided to launch satellites for intelligence and ordered related ministries and agencies to list their plans to make use of their satellites.[55] On 10 September 1998, PM Keizo Obuchi announced that Japan might launch its own reconnaissance satellite, to improve Japanese military capability and facilitate monitoring missile developments in North Korea. The main opposition leader, Naoto Kan, also endorsed the idea and a task force in the ruling LDP conducted a series of meetings with government officials and contractor representatives to develop an implementation plan. By 6 November 1998, the cabinet decided to develop and launch four IGS satellites with reconnaissance capabilities by 2002, citing security concerns over North Korea's rocket launch. The decision to launch the 'spy-sats' was also formally announced by Hiromu Nonaka, the chief cabinet secretary. The government established a panel at the cabinet secretariat to discuss the budgetary and technical requirements for the satellites. The committee headed by the deputy chief cabinet secretary Tejiro Furkawa, involved representatives of the Cabinet Information Research Office and the Cabinet Office of the National Security Affairs and Crisis management, the Foreign ministry, the Defence agency, the STA, the Ministry of International Trade and Industry, and the Post and Telecommunication Ministry were also represented on the committee. The satellite project according to Nonaka was slated for completion by the end of fiscal 2002 and the initial stage of the project was to receive around ¥10 billion (US$83 million) and its total cost was expected to be ¥150 billion. Japan's NASDA was expected to play a central role in the satellite development, with a proposed launch by March 2003. The total project had envisaged launch of eight spy-satellites through 2006 for reconnaissance over North Korea.[56]

On 21 December 1998, Prime Minister Keizo Obuchi's cabinet approved a draft budget for fiscal 1999 in which the STA was allocated 6.8 billion yen for the plan. The agency would be in charge of making and launching the satellites. The Cabinet Secretariat, which would be in charge of the entire system, was allocated 1.4 billion yen in the draft budget. Related ministries were directed to "work in close cooperation" because the plan involved a considerable amount of funds and systematic backup. The Self Defence Agency convinced the Ministry of Finance to allocate up to $1.7 billion for the programme. After gaining cabinet approval, Prime Minister Keizo Obuchi's government would include funding for a study into the project in the supplementary budget for the fiscal year to March 2000.[57]

Prime Minister Keizo Obuchi opened the 145th session of the Diet on 19 January 1999 with a speech in which he spelt out his views on the major issues facing the nation in the final year of the 20th century. The Prime Minister said "...in order to ensure the security of our nation in the international environment which surrounds us, I will take measures beginning with the introduction of information-gathering satellites in order to collect, analyse, and transmit information which can be of use in ensuring our national security and in managing crises".

Whether the satellites would be developed domestically or imported from the United States was initially undecided. The government wanted to develop the satellites domestically, while the ruling Liberal Democratic Party (LDP) wanted to import them to save time. In May 1999 the Japanese Government decided not to buy American satellites, but to have Japanese industry develop the satellites. According to Secretary Nonaka, the reasons were that faster reaction would be possible: NASDA has the necessary technologies, and 1 meter resolution imagery would provide significant information. The US government had unofficially asked Japan to buy a US-made satellite. The Japanese government may eventually have bought some US-made parts.[59]

By early 2000, Japanese impatience with its pacifist manifesto had reached a crescendo and a House of Representative research commission was established in January 2000 to revise its 'peace constitution' that occupying US forces had drafted nearly half a century ago. Revising or abolishing the war-renouncing Article 9 of the Constitution to enable the armed forces to execute the right to collective self-defence was the core issue of the commission's 700-page report that was submitted to the parliament in 2002.[59] This had significant implications on overall Japanese policy and by September 2003, press reports indicated that Japanese officials wanted to have both of their big military space projects—a satellite imaging system and a multi-tiered missile defence system—fully operational by the next two to three years in addition to a GPS augmentation system that could be used for military communications and, if required, missile targeting, by the end of the decade.[60] A brief description of the above and other Japanese space militarisation efforts follows.

Information Gathering Satellites, (IGS)

Most reports indicated that Japan would commence its space militarisation endeavours with deployment of a spy-satellite constellation for "information gathering" to boost its intelligence network. The Prime Minister's cabinet office was assigned the responsibility of procurement of the satellites from Mitsubishi Electric Co. (MELCO) and also for operational management of the constellation. The Mitsubishi group proposed detailed plans for a series of four IGS. Two of the satellites would have optical sensors with 1 metre resolution, and the other two would have radar imaging capabilities. The proposed satellites were meant to orbit at an altitude of 500 km using a large satellite bus based on a standard commercial Mitsubishi bus.

The IGS project originally intended to launch a constellation of eight satellites through 2006.[61] The full constellation was designed to give the country access to photographic images with a ground resolution of 1 metre and cloud-penetrating, phased array radar satellite data with a ground resolution of 1–3 metres. A six-satellite constellation was expected to be on orbit by the end of 2005 and by March 2009 another pair of improved satellites was slated for launch. A "third generation" of very capable satellites after 2010 were also conceptually planned as reported by Masato Nakamura, a CSIC researcher.[67]

Thus, finally on 28 March 2003, shrugging off its decades-long abstinence from space militarisation, Japan launched its IGS-1A and IGS-1B from *Tanegashima* launch site on an H-2A launch vehicle. These satellites were also designated Optical-1 (IGS-1A) and Radar-1 (IGS-1B) based on the payload carried. The IGS-1A utilises an optical camera with a black and white resolution of about 1 metre and IGS-1B uses Synthetic Aperture Radar (SAR) technology to see through clouds or darkness for night and low visibility observations. Speculations were rife that with its spacecraft in place and testing presumably under way, the pair would soon have the capability of imaging any location on the globe. The launch of the second pair of IGS satellites, Optical-2 and Radar-2 scheduled for 10 September 2003, by the Cabinet Office's IGS promotion council was postponed three times due to technical difficulties. When Japan finally attempted the launch in November 2003, it resulted in an H-2A rocket malfunction that destroyed both satellites. Since the satellites need to be rebuilt, the full four-satellite constellation is likely to be delayed until at least fiscal year 2006, and till then Japan's spy-satellite project is effectively crippled. Nevertheless, certain media reports aver that the Japanese government panel has approved plans to send two spy-satellites in orbit starting in 2005 at a cost of $635 million. The first of these probes, due for launch next year, is expected to carry cameras designed for taking high resolution pictures of ground targets while the second probe programmed for launch in 2006, would use radar to analyse topography.[63]

Quasi Zenith Satellite System, (QZSS)

Japan is also known to have mooted plans for development of a QZSS constellation of three Navigation Satellite (Nav Sats) and S-band mobile communications satellite capable of 25 centimetre positioning accuracy by 2008. The system is meant to supplement the use of the American GPS system and expected to comprise three Nav Satellites each equipped with communication, broadcasting and positioning capabilities. The quasi-zenith satellite is an artificial satellite of the satellite system where one satellite always stays near the zenith in Japan by positioning at least three satellites synchronously on the orbit inclined at 45 degrees from the geostationary orbit. As the ground surface orbit inclined at 45 degrees from the geostationary orbit. As the ground surface orbit draws the shape of number 8, it is also called "Number 8 orbit Satellite". It can obtain a high elevation angle to reduce the influence of buildings, steep slopes, etc. Japanese government officials and the executive Vice President of ASBC (the consortium that operates the QZSS and markets its services to the government and private sector) assert that the JDA proposes to use the constellation only for military communications and that no official approach has been received from the JDA for using the constellation for targeting precision-guided weapons.[64] However, the option for optimum military exploitation of the QZSS has not been foreclosed altogether. For example, Setsuko Aoki, professor at Keio University, Department of Policy and legal expert on Japan's space law opines that any future direct military use of space assets would make changing the resolution on space necessary, however, "since the QZSS would be largely used for commercial purposes, interpretation of non-military uses of outer space, Japanese style, seems still workable even if it is also used for military positioning and missile guidance".[65]

The foregoing also needs to be viewed in conjunction with media reports of Japanese plans to "link Japanese fighter jets to a US-developed satellite-guided bombing system beginning next year".[66] Taking Japanese proficiency at GPS receiver technology into consideration, the proliferation of GPS and GIS providers, and its own indigenous QZSS programme, it could safely be surmised that both the technology and requisite doctrines are mature; dormant/latent satellite navigation, targeting and positioning capabilities will soon be available and in the face of threats to its survival or interests, Japan's pacific countenance may dissipate generating temptation to make use of its latent/immediate capabilities for military purposes also rather than confining such potent capabilities to just road traffic and personnel locating applications.

Space Surveillance

The STA in the late 1990s had also announced plans to install a debris-tracking system in Okayama prefecture, western Japan. The system, which was expected to go into operation in 2002, was to consist of a new ¥0.8 billion ($5.9 billion) optical telescope and a ¥1.2 billion ($9 million) battery of phased array antennas which could coordinate to scan the low earth orbit region to a distance of approximately 1000 km. This Japanese system was meant to track objects of less than a metre in diameter at 1000 km and 1 metre sized objects up to 600 km. The facilities were to be under the jurisdiction of a Tokyo-based space forum with budgets from the STA.[67] While the system is apparently meant for benign civilian applications, it also endows military space surveillance capabilities which essentially comprise one of the primary building blocks of a 'counter-space' capability. While this above conjecture is inferential; it is not inconsequential, especially in view of Japan's emerging missile development programmes which are deliberated upon as follows.

Missile Developments

Japan is known to have been engaged in a tacit Ballistic Missile Defence (BMD) dialogue with the US since the early 1980s. The discussions were formalised in 1987 with both countries signing an "Agreement concerning Japan's participation in research for the Strategic Defence Initiative (SDI)". Soon thereafter, the US Department of Defense (DOD) sponsored the Western Missile Architecture (WESTPAC) studies (1989–1993) and conducted a series of research projects which were based on their mutual defence requirements generated by the proliferation of ballistic missile technology. The study, which cost $8 million and took four years to complete, examined the feasibility of defending the Western Pacific and Japan from North Korean ballistic missile attacks during the 2000–2005 period and concluded that the Nodong-1 was the major threat to Japan. It recommended that Japan adopt a satellite-based defence communications network; acquire Theatre High Altitude Area Defence (THAAD) as a "first-tier" BMD overlay; and examine the use of a sea-based BMD system.[68] Other studies conducted on the subject also led to recommendations for a two-tiered Japanese TMD architecture. The architecture was to combine the lower-tier Patriot Advanced Capability (PAC-3) and upper tier THAAD.[69]

The Iraqi use of short-range Scud ballistic missiles during the 1991 Persian Gulf War, coupled with the North Korean test of Nodong missile into the Sea of Japan in 1993 provided the Japanese government the impetus for official discussions with the US on a joint TMD programme. Thus, the bilateral Japan–US Theater Missile Defence Working Group (TMD-WG) was formed in 1994. Japanese concerns over regional missile proliferation were further increased in the mid-1990s with a week-long Chinese military exercise in March 1996, which included the launching of numerous Short Range Ballistic Missiles (SRBMs) near the Taiwan Strait on the eve of Taiwan's presidential elections, thus forcing the Japanese to consider the possibility that it might be threatened by Chinese intermediate range missiles in the context of a US–Chinese crisis over the future of Taiwan. North Korea's test of its 'Taepodong-1' missile in 1998 only served to precipitate matters and broaden support for BMD in Japan beyond defence analysts and experts. North Korea's Nodong-1 (Scud Model-D) Medium Range Ballistic Missile (MRBM) is arguably of greatest concern to Japan. It has a range of 1,000–13,000 km and could reach most of Japan. Secondly, the Nodong provides the core technology for the longer range, two-stage Taepodong. The Taepodong-1 has a range of 1,500–2,000 km and could reach all of Japan. The North Korean missile fired over Japan on 31 August 1998 was apparently a Taepodong-1 with a solid fuel three stage. Japan then had no means of protecting itself against ballistic missile launches and until it got its shield in place, the US had considered using its own Aegis warships armed with SM-3 missiles to protect Japan and American forces in Japan against ballistic missiles.

As of December 1998, Japan had officially decided to go ahead with joint research on the issue of missile defence and a Memorandum of Understanding (MoU) was signed with the US for research on Navy Theater Wide Defences (NTWD), an upper tier sea-based system of TMD on the rationale that it did not violate the Anti-Ballistic Missile (ABM) Treaty and was also not related to the controversial US National Missile Defence (NMD) System.[70]

Finally, by June 2003, Japan had mooted plans to develop advanced interceptor missiles to beef up its defence system amidst fears that North Korea had up to 170 medium-range missile units targeting Japan. Subsequent to the government "unofficially learning" that North Korea had 160–170 medium-range Rodong missile units targeting Japan and fears that North Korea might have acquired technology to reduce size of nuclear weapons for mounting on ballistic missiles, the Security Council of Japan and the cabinet adopted a plan to allocate funds for two types of missile systems; a Standard Missile-3 (SM-3) and a Patriot Advanced Capability-3 (PAC-3). The SM-3 is intended to be mounted on Aegis equipped destroyers to intercept ballistic missiles in outer space. The state of the art PAC-3 is designed to intercept missiles closer to the earth's surface during their descent prior to impact. The JDA currently possesses SM-2 systems mounted on four Aegis-equipped Maritime Self-Defence Force ships, as well as PAC-2 missiles for 27 launchers, including 24 anti-aircraft batteries operated by the Air Self Defence Forces. All the missiles, however, are designed to be anti-aircraft and are incapable of intercepting ballistic missiles which travel at much higher speed. SM-3s and PAC-3s are the latest version of the drastically improved SM-2s and PAC-2s developed by the US.[71]

Japan had decided on introducing an "Aegis destroyer-type" missile defence system, separate from the Japan–US joint research, because the US had decided to deploy the Aegis-based missile defence systems starting in 2004. JDA officials felt that while two Aegis-equipped destroyers were sufficient to intercept incoming missiles anywhere in Japan, a PAC-3 system deployed, for example, at Tokyo station would perhaps not be able to cope with a ballistic missile flying towards Yokohama station, about 30 km away. However, the PAC-3 scored over the Aegis because the Air Self Defence force had already introduced the PAC-2 and hence shifting to PAC-3 would not cause problems.[72] Japan's efforts to acquire a missile shield were premised on a major shift in its defence policy. During and even after the Cold War, the main threat was perceived to come from a large-scale invasion of Japan's homeland by the Soviet Union or China. Now the chief threat is seen to be terrorism and missile strikes, most immediately from North Korea. This was put forth for the first time in Japan's 2003 white paper on defence. The white paper also stated that protection against a ballistic missile attack was an important and pressing matter for national defence policy.

The Japanese government said on 19 December 2003 that it would buy from the US a two-stage defence system to shield Tokyo and perhaps other major cities from attack by North Korean missiles. The JDA expects to spend at least $4.2 billion to bring the system into initial operation by 2007 and full deployment by 2011, excluding maintenance and operating costs. The shield Japan will acquire is designed to detect a hostile missile as soon as it is launched. Japan will use SM-3 missiles on its four destroyers equipped with Aegis advanced electronic tracking, command and control system to intercept an incoming missile in its mid-course phase beyond the earth's atmosphere. If that first-stage defence fails, the PAC-3 will be launched from the ground in Japan to shoot down an incoming missile in the terminal stage of its flight. However, the PAC-3 interceptor has a mixed record of success.

On 19 December 2003, the government of Japan formally announced its decision to introduce a BMD system for Japan's protection. Based on this decision, it also sought to formulate a new national defence programme outline and a new mid-term defence programme by end of 2004. Details of the statement issued by the PMO have been placed as appendix 'J'.

The Japanese military, in keeping with its ballistic missile ambitions had on 31 August 2004 pressed for a 35% jump in spending on missile defence and intelligence systems for the next financial year. The military had asked for ¥144.2 billion ($1.3 billion) to build up Japan's US developed BMD system. The money was intended to be used mainly for buying sea-borne SM-3 missiles, upgrading the land-based PAC-3 anti missile systems and remodeling its Aegis destroyers. The Japanese Navy plans to conduct the first SM-3 tests in Hawaii by March 2008 to prepare for the operation of BMD systems. The overall budget demand of the JDA was placed at ¥4.93 trillion for the fiscal year starting April 2005.

Conclusion

The Japanese space (and space militarisation) programme are fairly well developed and by the year 2010, Japan would most likely have an advanced space capability boating of potent C4ISR capabilities, space-based navigation, positioning, targeting and a two-tiered BMD architecture in addition to the dual use capabilities its civilian space programme already affords. This coupled with its gradual conventional force modernisation programmes, its latent nuclear and ballistic missile capabilities would soon project it as a significant military force to contend with, not only in the Asian continent but in the entire world.

ISRAEL'S MILITARY SPACE PROGRAMME

Unlike most of his contemporaries who used conventional weapons like swords, spears, elephants, etc. to achieve the high ground, increase reach, and deliver their lethal blow, the diminutive David used the most unconventional weapon of his times—a 'sling', against a massive Goliath to deliver his lethal blow. This lesson of 'asymmetric warfare' has not been lost on the Israelis who till date continue with their quest for an 'asymmetric advantage' to contain the overwhelmingly numerical superiority of their adversaries. Thus in spite of possessing demonstrably powerful military capabilities in the region, it continues its quest for achieving the above-mentioned advantage so as to ensure its survival and also offset its disadvantage of lack of strategic depth. Hence Israel, has attempted to harness the tremendous advantages and opportunities afforded by space for its national security and in spite of financial constraints has made significant progress in its space endeavours.

Origin and Development

In 1960 the Israeli Academy of Science and Humanities established the National Committee for Space Research (NCSR), which paved the way for an Israeli space programme and official support for space research. However, NCSR was more geared towards space research and education rather than establishment of a space programme, and therefore apart from space research, the Israeli space programme at that time was non-existent. The only worthwhile achievement of the NCSR in the 1960s was the launch of a two-stage sounding rocket in 1961. Israel's missile programme also began in 1960 and on account of its threat perceptions, it was only natural that ballistic and other missile developments were accorded greater priority as compared to space.

Thus, as in the case of most other space programmes, Israeli Space Launch Vehicles (SLVs) also developed as an off-shoot of its Ballistic Missile (BM) programmes. There exists very little difference between Israel's Shavit (SLV) and Jericho (BM), both use the same family of rocket motors and most analysts on the subject opine that "the Jericho-2 is a Shavit minus the upper stage, which is replaced by a warhead". The interdependency of the Shavit and the Jericho reflect the blurred line between civilian and military developments in Israel. Firms involved in civilian space research also work on sensitive military projects, including nuclear and missile development. For example, Israel Aircraft Industries (IAI), the main contractor for the Shavit space rocket, also builds offensive and defensive missiles. Shavit's launch pad derived from military rather than civilian technologies betrays its military origins; Shavit is launched from a Transporter-Erector-Launcher (TEL), a platform used more often for military missiles rather than for satellite launchers.

Israel's small size and vulnerability to surprise attack have always resulted in its according high priority to intelligence and early warning. From time to time, the Israeli Defence Forces (IDF) reportedly received satellite imagery from the US, but the resolution was degraded, the coverage was limited, and the images were not provided in real-time. Since the two-front surprise attacks that began the 1973 Yom–Kippur war, Israeli intelligence agencies accorded high priority to developing independent sources of space-based intelligence. In fact, one of Israel's former Chiefs of Staff, Mordechai Gur had charged that immediately prior to the 1973 Yom–Kippur war, the US had withheld critical intelligence information, obtained by reconnaissance satellites, on Arab offensive military formations.[73] Israel thus became convinced of the need to have its own satellite capability after the US failed to provide it with satellite intelligence when it was most needed and thereafter the Israelis continued to perceive inconsistencies in US policies in supplying satellite imagery and information. The centrality of intelligence and early warning was emphasised as the countries in the region began to acquire ballistic missiles and weapons of mass destruction. Occasionally, the US shared strategic and space-based intelligence information with Israel. However, despite the close defence cooperation with the American government, Israel did not have routine access to real-time satellite intelligence data. A Defence Ministry official was quoted as saying, "For years we have been begging the Americans for more detailed pictures from their satellites and often got refusals–even when Iraqi Scud missiles were falling on Tel Aviv…" At times, the IDF has turned to other sources, including Russia, which reportedly sold hundreds of satellite pictures of Syria, Iran and Iraq for about $1 million, as part of a secret cooperation agreement.[74]

After the war Israeli intelligence acquisition problems increased manifold due to the limits imposed on aerial reconnaissance missions as a result of the separation-of-forces talks with Egypt and Syria. New instructions contained in the interim agreement required reconnaissance flights over Egypt and Syria to be authorised by the then Prime Minster Yitzhak Rabin (1974–1977). The above instructions became the norm and continued even after the Likud took over power in May 1977. This led to consultations between the Director of Military Intelligence (DMI), Major General Shlomo Gazit and Professor Yuval Neeman, special advisor to the Defence Minister Shimon Peres to resolve the conundrum because "Rabin's order created a situation in which air force planes had to make do with diagonal photographs, which are not as good".[75] The crisis intensified to such an extent that in 1979, Israel's then DMI, Major General Yeshoua Saguy told the Defence Minister Ezer Weiznan and Chief of Staff, Rafael Eitan that Israel's peace treaty with Egypt had created enormous problems for intelligence acquisition and the only solution was overhead photography satellites which would be able to bypass the political obstacles and take pictures of areas of interest without generating diplomatic and other problems. The demand, however, was met with little enthusiasm and much cynicism. Nevertheless, feasibility studies on the subject were ordered and $5 million of the MI's $500 million budget was earmarked for a feasibility study on the production of satellite launchers, satellite and telescopic cameras by the Israel Aircraft Industries (IAI), the Arms Development Authority (Rafael) and El-Op Electro-Optic industries. The three units were asked to complete their study within ten months.

Traditionally, a deliberate component of Israel's armament strategy has been to seek funds through military exports to financially support its own technological advances. In the 1970s, Iran was Israel's primary source of foreign funds through Israeli military exports and cooperative developments. However, by the time the results of the feasibility study commissioned by the MI were over, the revolution fomented by Ayatollah Khomeini in Iran and the fall of the Shah in February 1979, put an end to cooperation between Israel and Iran on the development of advanced weapon systems. This led to the under-employment of many Israeli defence industries and a shortage of funds to sustain its military programmes. By the end of 1980, following the completion of the preliminary stage of the feasibility study, Saguy asked PM Menachim Begin (who also held the defence portfolio) for the go-ahead to proceed to the next phase of the project. While the go-ahead was accorded, financial constraints were resolved by IAI tying up with South Africa for budgetary support. The security ties between the two countries began in the early 1970s and intensified through the following decades, with South Africa replacing Iran as the major client of Israeli weapons systems. Towards the end of the 1980s, Israel had some $2.3 billion worth of signed contracts with the South African Defence Ministry.

By the beginning of 1982, a recommendation entitled "Ofeq program" was submitted for developing an observation satellite. The programme included timetable and preplanning for a ground station, and laid out budget estimates and personnel requirements. The team also built a model of the satellite. The recommendation was to develop the satellite independently, without relying on foreign know-how, so as to avoid having to rely on foreign sources and to enable flexibility and creativity. It was decided that Rafael would serve as the chief contractor for the development of the satellite. The launcher would be developed by Malam. The two large engines of the Shavit rocket would be built at Israel Military Industry (IMI)'s Giveon plant, while a third engine would be developed by Rafael.[76]

At the end of 1982, it was decided during a closed door meeting to establish an Israeli space agency. The principal decision makers were the PM Menachim Begin, Defence Minister Ariel Sharon and the former Director General of the Defence Ministry, Brigadier General (res) Aharon Beit Halahmi. The principal underlying purpose for creation of the agency was to pursue the programme to develop the Ofeq (Horizon) surveillance satellites and Shavit (Comet) satellite launchers. The brain behind the programme was Professor Haim Eshed, of the Asher Space Research Centre at the Technion Israel Institute of Technology, in Haifa. Professor Eshed was director of the space agency's project and is generally considered the father of Israel's space programme.[77]

Finally, in January 1983, the Israeli government authorised Professor Yuval Neeman (then minister of science) to establish an Israeli Space Agency (ISA) focussed on advancing Israel's space programmes unlike the NCSR which was focussed on space studies. In July 1983, the ISA was officially founded in Tel Aviv and became part of the ministry of science and technology; however, it received its initial funding from a special budget set up through the MoD. According to Haim Eshed, Israel's initial investment in its space programmes was driven by its strategic considerations, especially the ability to observe the activities of other states without violating international law. It is for this reason that the primary focus of Israel's space efforts has been and continues to be the development of high-resolution imaging capabilities. The ISA develops Israeli space policy, coordinates the national space development and international cooperation, supports applied and theoretical studies and helps Israeli industry in developing and marketing space-related products. Additionally, the NCSR (now under the Israeli academy of sciences and humanities), the Interdisciplinary Centre for Technological Analysis and Forecasting of the Tel Aviv University, the Asher Space Research Institute and other government-backed industrial consortiums like IAI, Rafael, El-Op etc. also assist, augment and promote Israel's space efforts.

Ever since its inception, ISA was known to collaborate with European and American agencies and by 1986, America's NASA and ISA had signed a general agreement for exchanging scientific and technical information. However, the US shuttle *Challenger*'s crash in 1986 had a tremendous impact on Israeli space endeavours (like in the case of the Chinese and Japanese) and affected the ISA profoundly in the following two ways.

- It caused the entire Israeli space programme to stop (the only plan to survive was the 'Hornet experiment').

- It drove home Israel's requirement of an independent launch capability, thereby pushing the Shavit programme.

The debacle, nevertheless was temporary and on 19 September 1988, launched its first satellite Ofeq-1 on a three-stage Shavit from its Negev desert launch pad. This was followed by other surveillance satellites in the Ofeq series, communications satellites like the AMOS, EROS for

Constituents of the Israeli Space Programme

Launch Vehicles	Versions
Shavit	Shavit (RSA-3)
	Shavit-1 (LK-A)
	Shavit-2 (LK-1)
	Shavit-3 (LK-2)

Satellites			
Type	Version	Launch	Remarks
Ofeq (IMINT)	Ofeq-1	19 September 1988	Experimental Test
	Ofeq-2	03 April 1990	satellites
	Ofeq-3	05 April 1995	Successful IMINT
	Ofeq-4	January 1998	Failed
	Ofeq-5	28 May 2002	Successful IMINT
	Ofeq-6	06 September 2004	Failed
AMOS (Communications)	AMOS-1	May 1996	Dual use
	AMOS-2	27 December 2002	MilComm Payload
EROS (Commercial observation)	EROS-A1	05 December 2000	Dual use

Figure 6.2

Earth-observation, etc. A brief glimpse of the primary constituents of the Israeli space programme is available in Figure 6.2. It is evident that though Israel doesn't have an elaborate programme at the moment, it is an ambitious endeavour and likely to see some more significant progressions in the near future.

Launch Endeavours

Israel began development of guided missiles in the 1950s. However, these efforts were inclined towards military missile developments and not towards rocket development aimed at space launches. Israel commissioned its *Jericho-1* ballistic missiles from the French[78] firm Dassault aviation with the basic purpose of possessing a surface-to-surface BM system capable of delivering a 750-kg warhead with a range of 235–500 km and a CEP of less than 1km. By the early 1970s, Israel had started indigenously constructing the Jericho missiles and in October 1971, the *New York Times* reported that Israel's Jericho was 'nuclear capable' with a range of 300 miles and a 1000–1500 lbs payload. As borne out earlier, Israel's early rocketry developments were aimed at ensuring its survival in a hostile environment and dissuading adversaries from contemplating an all-out effort to bomb Israeli cities and destroy Israel itself. In the 1970s and early 1980s, qualitative improvements on the Jericho continued and using its Jericho BM technology, IAI developed and designed the Shavit SLV to place small satellites into Low Earth Orbit (LEO). Finally, on 19 September 1988, Israel placed its first satellite into orbit using a Shavit SLV designed and manufactured by IAI. Shavit was a 3-stage, solid propellant SLV designed to carry a 250-kg payload into space. Using the orbital parameters of the satellite launched, the US Lawrence Livermore National Laboratory concluded that the Shavit could be reconfigured as a BM capable of delivering a 500 kg warhead up to a range of 7,500 km. The following year, in 1989, the two stage, solid propellant Jericho-2 entered into service. Media speculation was rife that configured as a missile, the powerful Shavit rocket could hit every capital as well as city in Europe, Russia and even China. US officials were known to aver that the first two stages of the Shavit consisted of Israel's two-stage *Jericho-2* nuclear missile, "the *Jericho-2* is a Shavit minus the upper stage, which is replaced by a warhead".[79] Additionally, it was also speculated that judging from the power of the Shavit and its retrograde westerly launch, a *Jericho-2* consisting of Shavit's first two stages would be able to fly up to 4,500 km with a one-ton payload. Both Jericho and Shavit also use the same family of rocket motors manufactured by TAAS (formerly Israel military industries). Shavit's mobile launching pad also betrayed its military origins. Finally, in June 2001, the speculation was laid to rest with an Israeli official confirming that the "Shavit is Jericho".[80]

It would be imperative to note that due to its geographical location and geo-political considerations, Israel is the only country that launches its satellites westwards, against Earth's rotation, in order to avoid endangering civilian populations and also to prevent the over flight of neighbouring Arab nations. This requires a sacrifice in payload capacity so that more fuel can be added to provide the necessary thrust during launch. Israel's launching facility is located near the Palmachim Air Force Base, close to the Mediterranean coast and south of Tel Aviv. In 2002, Israel initiated negotiations with Brazil to use the Alcantara launch site for the Leo link line, although no agreement has been finalised. Launches from Alacantara would allow Israel to launch satellites not only eastward, but also near the equator where the Earth's rotation provides maximum boost.[81]

The developers of the Shavit, IAI/MLM (systems division) had also planned a commercial Shavit upgrade called "Next". However, this name is no longer used and the proposed upgrade configuration is now called Shavit-2. In pursuit of enhanced launch capabilities, IAI in May 1998 had announced an agreement with the Coleman Research Corporation to collaborate in the manufacture of small expendable satellite launch vehicles using technology developed for the Shavit. In addition, IAI/MLM had also initiated a LEO Link programme to market launch services using current Shavit-1 (LK-A) launchers while developing the LK-1 and LK-2, its next-generation launchers. The following were the capabilities envisaged.

- LK-A — For 350 kg-class satellites in 240x600 km elliptical polar orbits.
- LK-1 — For 350 kg-class satellites in 700 km circular polar orbits.
- LK-2 — For 800 kg-class satellites in 700 km circular polar orbits.

A Shavit LK air-launched satellite launcher is also being proposed by

ISA and IAI. The booster would be a standard Shavit-1 without a first stage that would be dropped from a C-130 aircraft.[82] To help cut costs and improve its ability to put satellites into orbit, Israel is eyeing plans to use aircraft and Pegasus type launch vehicles to develop an inexpensive, fast-responsive capability. Chaim Ehsed, Israel's head of the defence ministry's space programmes said in 2003 that Israeli researchers believe that they can launch a 220-lbs satellite from an F-15 within the next five years.[83] Scientists at RAFAEL, Israel's Armament Development Authority, are also examining technology to launch satellites from F-15 fighter jets. The concept is to upgrade the Black Sparrow missile (used for testing the Arrow Weapons System) with a more powerful engine and install a microsatellite in its nose. Israel expects to have this technology available in 2008. In an interview on 30 July 2003, Chaim Eshed said that he envisioned, perhaps within five years, that the Israeli Air Force would be able to launch multiple satellites on demand from fighter aircraft that would range in weight between tens of kilograms to less than 100 kg.[84]

Anti-Missile Systems

In 1986, the US and Israel had signed a Memorandum of Understanding (MoU) for joint development of an Arrow Anti-Tactical Ballistic Missile (ATBM) system. The Arrow programme begun as part of the US SDI (Strategic Defence Initiative) is a key component of Israel's programme to develop an anti-missile shield programme linked to Israeli military surveillance satellites. Israel conducted the first test of the Arrow-1, designed as a theatre defence missile system capable of intercepting SS-1 Scud missiles, SS-21 missiles, Iraqi Al-Hussein, Chinese CSS-2 missiles, etc. In spite of the first test recording a failure, later tests recorded varied degrees of successes. Arrow-2, an operational version of Arrow-1 followed in 1995 and was largely successful in its first test. In October 2000, the first batteries of the Arrow-2 system were declared operational and by December 2003, Israel had conducted up to eleven interceptor tests and six tests of the complete system. As late as July and August 2004, Arrow's mixed bouts of successes and failures continued with an Arrow-2 successfully intercepting a confiscated Iraqi scud in July as against a failed simulated interception in August. Incremental progressions on the system are likely to continue till the system is operationalised.

Israel is also known to be working on a system designed to destroy incoming short-range rockets with a concentrated laser beam. The project known as Tactical High Energy Laser (THEL) is a US–Israel joint enterprise still in a developmental stage and its future remain uncertain. However, its mobile version the MTHEL is advancing rapidly and is expected to provide a complete prototype by 2007. MTHEL represents the first mobile-directed energy weapon that will be able to destroy tactical airborne threats in midair such as rockets, missiles, artillery shells, etc. The target destruction is achieved by projecting a highly focussed, powerful laser beam, delivered by a chemical laser with enough energy to affect the target and explode it in midair. This operational concept is offering the first "re-usable" interception element. Existing interceptors use kinetic energy kill vehicles (such as fragmentation warheads) which are not re-usable.

Satellites

Ofeq

Israel's Ofeq space surveillance programme is an elemental component of its multi-layered 'Homa' (wall) anti-missile defence shield aimed at affording protection against hostile BM attacks. The Ofeq would be Israel's first layer of the defence shield, designed to spot incoming threats and alert defence systems like the Arrow-2 interception systems. Apart from providing information on BM launches and other related information to its Arrow system, it is also designed to provide high resolution imagery for military purposes like detection, surveillance and other routine observations ranging from discerning illegal settlements in the occupied territories to keeping track of nuclear developments in its hostile neighbourhood.

On 19 September 1988, with its launch of an *Ofeq 1* test satellite on a Shavit launcher, Israel established itself as the eighth space power in the world and the fourth space power in Asia.[85] This 156-kg satellite was reported to be a test vehicle designed to lead to the development of an orbital reconnaissance capability and it re-entered Earth's atmosphere in January 1989. To avoid flying over Syria or other Arab countries, a highly unusual flight-path was used which headed north-west into the Mediterranean placing the satellite in a retrograde orbit at an inclination of 143.35. Ofeq's orbit limited the satellite's view to areas 37° north and south of the equator. *Ofeq 2* was similar in weight and technical characteristics to *Ofeq 1.* It was launched in April 1990 and had an orbital lifetime of 3 months, both were spin stabilised. *Ofeq-2,* additionally offered the possibility of two-way communication (*Ofeq-1* could broadcast information only). *Ofeq 1* and *2* were both experimental models.[86]

This was followed by its first operational imaging and Early-Warning (EW) satellite, the *Ofeq-3* launched on 5 April 1995, which weighed 225 kg at launch including a 36-kg payload. *Ofeq-3* was primarily an imaging satellite with ultra-violet and visible imaging sensors and a resolution of 1.8 metres. Like the previous Ofeqs, *Ofeq-3* was also positioned at an orbital inclination of 143.35, thereby covering its Arab neighbours like Syria, Iran, Iraq, etc.

However, Israel's space programme suffered a severe blow in January 1998 with the failed deployment of its next military surveillance satellite, the *Ofeq-4.* The *Ofeq-4* was slated to replace *Ofeq-3,* which had already exceeded its anticipated life-span. The *Ofeq-4* programme was an ambitious project and several of its top defence companies were involved: Elisra built the video compressor, ELOP manufactured its cameras, Elbit supplied the computer, Rafael power and electricity, etc. This led Jane's defence weekly to report in 1998 that "the Offeq-4 failure means that there will be a significant gap in the satellite series that was designed to give Israel a surveillance capability independent of the USA". Ever since thirty-nine Iraqi Scud missiles slammed into Israel during the 1991 Persian Gulf War, the US and Israel had a defence pact signed which was aimed at providing the Israeli's immediate warning in case of BM launches against them. Based on the pact, American EW satellites were to give warning on BM launches providing the Israel's about seven minutes to get into shelters.[87] However, Israeli perceptions on American inconsistencies prevailed and subsequent to Iran test-firing its nuclear capable *Shahab-3* missile on 5 May 2002. *Ofeq-5* was launched on 28 May 2002. *Ofeq-5* filled a yearlong gap in Israeli intelligence gathering caused by the *Ofeq-4* failure. The military, nevertheless, did manage to extend the life of *Ofeq-3* from its planned three to six years, but it burned up in the atmosphere by 2001. Since then, the military had been purchasing services from its commercial satellite 'EROS' (Earth Resource Observation Satellite).

The *Ofeq-5* is a three-axis stabilised satellite, designed by IAI/MBT division. Its design is based on proven technologies employed in the

Ofeq satellites and latest developments using leading edge technology. The satellite encircles the earth in an elliptical orbit from east to west, every hour and half, at an approximate angle of inclination of 143 degrees, taking it over Iran, Iraq, Syria, etc. once every 90 minutes. *Ofeq-5* is designed to operate at altitudes between 370 and 600 km above the surface of the earth. Its projected lifetime is approximately four years. *Ofeq-5* belongs to the class of small and lightweight satellites. *Ofeq-5's* light weight allows for maximum agility over target to yield rapid image acquisition. It acquires images in swathes ahead of satellite trajectory, beneath it and lateral to it. It is also capable of 'orbital detuning' in response to ground commands which means that inclination orbits can be adjusted in response to special war contingencies in order to bring them over requisite areas of interest.[88] It was also reported to have high-resolution telescopic cameras designed by Elbit systems which could produce imagery of objects as small as 1 metre across from an altitude of 600 km. In fact, Israel's defence minister Benjamin Ben-Elizer was euphoric about *Ofeq-5's* high resolution capability and claimed that the satellite had detected objects as small as half-a-metre from an orbiting altitude of 500 km.[89] However, the euphoria is not all-pervasive and specialists in strategic studies like Reuven Pedatzur of Tel Aviv University are more realistic and opine that unless Israel has eight satellites with two over Iraq and Iran, there is nothing Israel can get which the US doesn't already provide.[90]

In view of *Ofeq-5*'s inability to adequately surveill its area of interest (approximately 879,730 km), and Iran's impending launch of *Shahab-5* amongst a host of other factors, Israel launched its *Ofeq-6* on 6 September 2004. Following a great deal of preparation and secrecy, a Shavit missile was launched from the Air Force firing range at Palamachin, carrying the *Ofeq-6* on its nose. After successful completion of its first two stages, the third stage malfunctioned and sent the satellite plummeting into the eastern Mediterranean Sea near the city of Ashdod. The failed launch cost the Israeli exchequer some $50 million ($100 million as per *Haaretz News),* but did not lead to a virtual collapse of its surveillance programme due to the presence of *Ofeq-5* and its commercial EROS satellite. Notwithstanding, the above loss is a serious dent in Israel's surveillance and consequently deterrence and other capabilities at the most inopportune of times. Media reports were rife with stories that the inadequacy of a single satellite in orbit was apparently manifest in 2003, when Iran clandestinely dispersed its fifteen known nuclear installations across the country (an area of approx. 636,000 square miles) and *Ofeq-3* in spite of American support was unable to locate the sites.[91] *Ofeq-5,* on its own, without a partner, was found to be incapable of gathering all the data Israeli intelligence needed to fully appreciate the intentions of Iran's military leaders. The defence establishment, shaken up by the failed launch has set up a special team to investigate the reasons for the failure and also plans to push forward with more launches to boost its surveillance over Iran's nuclear weapons programme.[93]

Early warning, surveillance and real-time reconnaissance have always been of major importance to Israel's defence planners in offsetting the threat to national survival posed by the massive conventional forces of the neighbouring states. Overall, the Israeli military would be satisfied with a minimum constellation of at least threat three to four satellites which would comprehensively cover its areas of interest. The *Ofeq-6* was designed to provide another layer in Israel's missile warning system and real-time data on Iran's missile programme, but failed. However, Israel has another *Ofeq-7* slated to be in place by 2007–08. In addition to the above electro-optical satellites, Israel's advanced radar satellite; TechSAR is also being readied by the MoD for some time in 2005–06 and would expand Israel's imaging options in all weather conditions and night. In response to the military's needs, Israel's MoD aims to exploit the full spectrum of imaging capabilities, including increasingly high-resolution electro-optics, infra-red, hyper spectral SAR and three-dimensional imaging. "The idea is to create constellations of many satellites in a wide spectrum of wavelengths", Brigadier General (Retd) Chaim Ehsed was quoted as saying on the vision for future Israeli programmes.[93] Media reports in 2003 also indicated that Israeli imagery resolution capabilities had progressed beyond 10 centimetres and Israel was getting close to the so-called "diffraction limit", where light is diffracted to a point where pictures cannot be captured.[94]

Earth Resources Observation Satellites (EROS)

Apart from the Ofeq programme, Israel also makes use of its EROS dual-use satellites for complementing its observation capabilities. ImageSat International, in partnership with IAI, produces the EROS series remote sensing, dual-use satellites for which the defence ministry is the primary client.[95] IAI's first proposal for the EROS programme was submitted in 1993, and by 1996 IAI had concluded agreements with the Core Software Technology (CST) California, for development of the EROS system. Although press reports claimed that the first EROS launch was scheduled for 1997, the same was delayed because the Israeli MoD refused to approve the license. By 1988, a programme for a constellation of eight commercial LEO high-resolution imaging satellites was completed. This system was designed to provide potential customers including the Israeli military, with very frequent coverage of any point on the globe, compared to the more sporadic coverage available from a system of one or two satellites. The satellites were to be launched into LEO's at altitudes between 480–600 km using Russian launchers since Israeli launch capability was insufficient for both the altitude and weight of the EROS payload. In 1999, West Indian Space and the Israeli government had reportedly signed an agreement providing exclusive access to images of the Middle East region obtained by the first three EROS satellites for eight years.[96] On 5 December 2000, EROS-A1 was launched from a Russian Start-1 rocket from the Svobodny cosmodrome in Siberia. Weighing 250 kg, the small sat design is based on technologies developed for the Ofeq series of Israeli military satellites. EROS-A1 has a Panchromatic CCD (Charge Coupled Device) camera with a standard resolution of 1.8 m. However, by using the satellite in the over-sampling mode—when its camera is pointing directly down towards Earth—the resolution can be boosted to 1 m. Follow-on satellites in the series are expected to produce images with 0.82-meter resolution. Six satellites in the subsequent EROS-B class will be equipped with cameras capable of a resolution of 0.82 m, eve under poor lighting conditions.

AMOS

With regard to communication satellites, Israel developed the AMOS satellite series. *AMOS-1* satellite developed under the project was entirely an IAI project; it was launched in May 1996 on a French Ariane-5 launcher and provided communication services to Israeli and European customers. Although a technological success, the state comptroller called it an economic failure.[97] This was followed by a second generation dual-purpose *AMOS-2* launched from Kazakhstan aboard a Russian Soyuz rocket on 27 December 2003 for military and civil communications. *AMOS-2* is Israel's second commercial communication satellite and is positioned some 4 km from its predecessor, *AMOS-1,* which is expected to continue providing services till 2008. Speculation in the Arab media was rife that since AMOS had its orbit adjacent to the Arab Satellite Organisation's Arab Sat series which carries signals between the Arab states, one of AMOS's functions was to monitor Arab Sat traffic. Israel is also known to have plans for another more powerful military communications satellite, larger and more capable than the *AMOS-2* to be launched in late 2004.[98]

Micro Satellites and Other Programmes

Israel is also known to be considerably proficient in areas of satellite miniaturisation, electric boosters for satellite trajectory modification, seismic application of navigation, etc. Students of the Israel Institute of Technology (Technion) led by Dr Moshe Guelman have achieved considerable levels of specialisation in micro and nano satellite technology and attempted a 52-kg *Gerwin-1* techsat launch, which though a failure yields immense academic knowledge. However, it successfully launched its *Gerwin-2* techsat in July 1998. Israel is currently interested in developing what is considered the next generation of micro and nano-sats that weigh less than 100 and 10 kg respectively.[79] Israeli defence planners were known to have decided to focus on rapid development of a long line of micro and nano-sats to boost their space capability and offset the effect of shrinking budgets. According to Chaim Ehsed,

> As they did earlier in the nations UAV programmes, Israeli officials are focussed on simple cheap systems that can be launched quickly and that don't depend on the approval of another country. The Israeli defence ministry is working on a strategy to make micro and mini-sats and eventually even smaller nano-sats more capable. One promising option would be to split a large payload among several satellites. The satellites arrangement or formation would depend on the task. COMINT and SIGINT satellites would likely be arranged in long lines to quickly triangulate the position of transmitters of interest. Satellites could be flown in a cluster-circular, chevrons or a square to create a single, large virtual aperture of much greater resolution.[100]

Israel is also contemplating development of very small nano-sats for disabling satellites.[101] In fact, the Israeli Air Force has examined the operational need for a small micro satellite for every commander that could be stored at IAF bases and then be launched at short notice, if necessary. However, funding costs in June 2004 have reduced the budget for this programme to almost nothing.[102] Like China and India, Israel has also invested in Galileo, the European global navigational satellite system project. Israel had signed an agreement for the same in 2004.

Doctrinal Drivers

Israel is a country born in conflict, and its basic security situation has not deviated from that. Survival of the state is an all-pervasive national issue, and the short history of the state has been punctuated by a sequence of serious wars. Israel's survival, therefore, depends on its acquisition of 'asymmetric techniques' to offset its inherent disadvantages as also the numerical advantages of its adversaries. As the principal threat has shifted from the proximate land forces of Syria and Egypt to the longer-range missile capabilities of Syria, Iran and Iraq, Israel sees the need to upgrade and modernise her forces across the board. This need especially includes extending and upgrading the lifetimes of its weapons platform, developing over-the horizon strike capabilities, space, and ballistic missile defence. In response to the increased militarisation and acquisition of sophisticated military technologies by its adversaries and its own unique needs for survival, Israel has developed a "new war theory" based on which its future missions and force structures of IDF are being developed. The following are its main elements.[103]

- Ensuring that the element of surprise is utilised in any future conflict to achieve a decisive strategic victory for Israel.

- Developing deterrence policies and capabilities, particularly the "deterrence by doubt" factor which aims at intimidating decision makers in countries that might have tense relations with Israel. This type of deterrence would restrict the movements of Israel's opponents and limit their options and reactions as a result of their doubts about the size and type of Tel Aviv's potential response to any military operations.

- Maintaining the technological gap in the fields of armament and military industrialisation between Israel and the Arabs and making this gap as apparent as possible to frustrate the Arabs and convince them that it is impossible to defeat Israel in a future war.

- Possessing strategic weapons systems that can constitute an effective deterrent capable of reaching any target in the Arab region and the Middle East in general.

- Constant readiness to launch a military retaliation whenever necessary that wages total war on an opponent to achieve major objectives and stabilise the situation for long periods of time. A corollary to this readiness, as circumstances require, is to launch a limited retaliatory strike on a small scale that would not jeopardise the peace process.

In view of the foregoing, it may be safely surmised that the prevailing broad strategy for Israel would be to initially construct a comprehensive surveillance network of small spy-satellites in LEO to avoid being surprised and at the same time maintain its own 'element of surprise' capability. Its missile defence systems would provide it with its much-needed deterrence capability and the entire military edifice would be supported by its proposed space communications system. Israel's strategy of using small satellites coupled with its proficiency at orbital detuning as also micro and nano-sat technology would provide it the capability to fulfil its force-enhancement and space-support missions, in addition to an incidental 'ASAT' capability, if the need arose. Israel fully comprehends the capabilities and opportunities offered by space for enhancing its ability to achieve its security objectives. Israel's strategic doctrine dictates that space-driven RMA is critical to its future military operations. Israeli planners view space as a natural extension of air-power to be harnessed synergistically to its advantage as the balance of power in the region shifts increasingly to ballistic missiles.

Conclusion

Regular Israeli pronouncements on the subject like "the ability to utilise space is the cornerstone that supports air power", "Whoever cannot achieve this...will not be in the forefront of modernisation", etc. by former Israeli Air Force commander Major General Eitan Ben-Eliahu, an ardent space advocate as also pronouncements by its present commander Major General Dani Haloutz that "In summary, air and space power complement each other operationally through command and control of air missions, targeting, intercepting ballistic missiles, providing weather information over long-distance areas, and intelligence mission escorting. Building up of military forces should be based on exploring and understanding the trade-offs between air and space requirements—each can be operationally integrated to provide greater capabilities than perhaps might be available separately", serve to validate the view that Israel fully comprehends the full import of air, space and aerospace-power and once its finances stabilise, it would make all-out efforts to boost its comprehensive military power with a generous element of space-based capabilities.

INDIA'S MILITARY SPACE PROGRAMME

Background

Few would be aware of the fact that the roots of the Chinese Dragon's prowess in martial arts stems basically from India. The Chinese were taught the art of self-defence by "Bodhidharma", ("*Damo*" in Chinese), an Indian Buddhist monk who migrated to China from Kerala in southern India to proselytize his faith.[104] While *'Kalari-Payat'*, the precursor of present-day martial arts like 'Kung-fu', 'Karate', etc. is almost foregotten in its place of origin, the Chinese adopted it, mixed and mingled it with local know-how and tradition and perfected it into a lethal fine art. i.e. the Dragon perennially assessed his changing environs, changed and adapted with times, honed his survival skills and secured his realm; while the Tiger lay back, admired his stripes, licked his whiskers, crouched and occasionally let out a menacing growl to re-establish his authority in his shrinking environs.

This analogy best serves to illustrate the attitudinal divergence which reflects the promotion of military space programmes in India and China. While the Chinese, ever-sentinent of survival and threat have an advanced space militarisation (in fact weaponisation) programme and are quietly progressing towards 'Space-Control' missions by developing parasitic ASATs (Anti Satellite), etc. the Indians are yet to embark on a serious space militarisation or at least space security (survival of space assets) programme. Indian 'Pakistan-centric' myopia continues to prevail and since the Pakistani space programme is diminutive, India pays little attention to securing and ensuring survival of assets in space.

Both India and China are ancient civilisations bearing their own individual cultural, intellectual baggages and survival skills. Both have a civilisational heritage of intellectual and scientific pursuit in spite of repetitive threats to survival amidst hostile environs and invasions. Hence, while the astounding progress in space by both countries is not surprising, the overwhelming emphasis of one on space militarisation and the absolute lack of emphasis by the other on militarisation or for that matter even on space security (securing its assets in space) is truly amazing.

Development

India's remarkable space programme is conspicuous on account of the fact that even the Chinese, who are normally skeptical and cynical of Indian achievements are downright adulatory about it, claiming that "though the Indian space program has a history of just thirty years; it shows its stamina as a long-distance runner in space technology. In recent years, especially, it made great achievements and continually rewrote its records on an annual basis".[105]

In an organisational sense, the Indian space programme began in 1962 with the creation of the Indian National Committee for Space Research under the Department of Atomic Energy. The formation of ISRO in 1969 highlighted the transition of the space effort from a scientific undertaking of limited magnitude to a coordinated programme with specific goals and time-based projects in space applications and technology. The programme's organisational structure was further augmented in 1972 with the establishment of a space commission and a Department of Space.

India started from scratch, steadily growing to its present status in spite of technology transfer restrictions, etc. in a manner which was largely independent and autonomous, on account of its policy of non-alignment. In sharp contrast to the Chinese and even American and Russian space programmes, which were basically derivatives of their ballistic missile programmes, the Indian space programme had strictly civilian beginnings and its IGDMP (Integrated Guided Missile Development Program) was a later military derivative of its civilian space programme. Additionally, unlike the Chinese, American or Russian space programmes which are invariably interlinked with military programmes with the military having a dominant role in space affairs; in the Indian context the military has no say in space affairs and does not even find a slot for itself on the Indian organisational space chart.

Indian space efforts started with the establishment of the TERLS (Thumba Equatorial Rocket Launch Station) near Trivandum (now

Thiruvananthapuram) in 1963. The first phase of its space programme mainly involved creating infrastructure in the 1960s and space research was limited to sounding rockets. In the second phase during the 1970s, India undertook a series of experimental satellite missions, which were as below.

(a) Aryabhatta

The first Indian satellite built was launched into space with Soviet help on 19 April 1975.

(b) Bhaskara 1&2

Experimental earth observation satellites launched in 1979 and 1981 respectively.

(b) Rohini Series

The Rohini series of satellites were launched by India's own launch vehicles between 1979 and 1983.

(d) Apple

Experimental communicational satellite launched by an Ariane vehicle in 1981.

India's successful placement of *Rohini-1* into near-Earth orbit, on its own independently developed launch vehicle in its second attempt itself made the world take notice and also made India the sixth country in the world with independent launch capabilities. The magnitude of this achievement can be gauged from the fact that it took even a technologically advanced country like Japan nothing less than five attempts to put a satellite into space.

With the successful launch of *Aryabhatta,* India demonstrated its ability to design and fabricate orbital satellites on its own. With the exception of the solar panels and batteries, all of *Aryabhatta's* systems were manufactured in India.

Since the 1980s, India with its own indigenous space technology has launched the INSAT series for telecommunication and other applications. With the launch of the IRS (Indian Remote Sensing) series, it became the fifth country in the world to develop and launch remote-sensing satellites.

Its IRS series provides data in a variety of spatial resolutions starting from 360 metres to an extremely high resolution of 1 metre. The IRS TES (Technological Experimentation Satellite) is also known to be capable of providing less than one metre of resolution.

India's launch abilities have also been consistent with its exemplary performance; with its launch of GSLV (Geo synchronous launch vehicle) in 2001, it became one of the select six countries to reach the geosynchronous orbit. India has launched about 36 satellites so far, of which 17 were launched using Indian launch vehicles. To date, India has developed a complete space system. It has mastered the technology to manufacture and launch carrier rockets and satellites as well as that of telemetry, tracking and recovery. It also has a large contingent of manpower and generally does not lack research funds.

India's Military Space Programme

While discussing Indian military space developments, due credence must be given to the fact that India consciously chose not to pursue a military space programme in spite of being aware of the immense "force multiplication" capability space would have bestowed on its conventional military might. For instance, way back in the early 1970s, six months prior to the 'Bangladesh' war the IAF (Indian Air Force) had issued an ASR (Air Staff Requirement) setting out the qualitative requirements of an Indian military reconnaissance satellite, at a time when only aircraft-based special colour film cameras were being tried out by ISRO for civil remote-sensing experiments.[106] However, no such capability was afforded to the IAF and the IAF had to make good with its conventional resources. Even in the early 1980s Indian military writing on space was profound and a senior IAF officer had authored writing on techniques of avoiding detection by enemy military satellites which again was remarkably prescient in its details.[107] Certain Western analysts also aver that "perhaps no developing country has more of an appreciation for military space than India. Indian defence professionals tout the vital contributions of spacecraft to intelligence acquisition, battle management, and weapon precision, and they understand such technologies to be a sine qua non for all nations seeking power status. We should look at the orientation of India's space-minded military and space-aware officials as indication of the likely global trend".[108] Hence it may be safely inferred that what probably restricted a country with the doctrinal inclinations and technological capabilities from an active space militarisation campaign was lack of political and institutional conviction or endorsement and partly its penchant for inter-services turf battles which is in no small measure responsible for its Aerospace Command project not taking off.

Apart from its IGMDP (Integrated Missile Development Programme), which is a military derivative of its space launch capabilities, India's military utilisation of its enormous space capabilities is miniscule and military participation is non-existent. The primary benefit has been that the solid-rocket motor and control, guidance and navigation technology developed for the SLV-3 and ASLV boosters have been adapted for India's nuclear capable Agni range of surface-to-surface missiles under the previously-mentioned IGMDP. Similarly, Indian plans to develop ICBMs (Inter Continental Ballistic Missiles) may also benefit from PSLV (Polar Satellite Launch Vehicle) and GSLV (Geosynchronous Satellite Launch Vehicle) technologies.[109]

Amongst application satellites, the IRS (TES) launched in October 2001 affords military access to imagery of one metre resolution, which is of immense tactical and strategic importance. This is the only operational satellite system of exceptional value to India's armed forces. However, it is also grossly inadequate for Indian military requirements which is apparent from the fact that India also acquires imagery from America's commercial Space Imaging firm and has also approached Israel with a request for *Ofeq-5* data to supplement the TES imagery.[110] The lack of institutional support and the extent of military and space disconnect can be gauged from the fact that in spite of India's civilian DOS (Department of Space) having the largest constellation of remote-sensing satellites in the world, the Indian military goes around scouting for commercial imagery from Space Imaging, Israel's Ofeq, etc.

The military primarily uses space for surveillance and reconnaissance and that too in a very limited manner, basically as an auxiliary to its ground and air-based surveillance capabilities.

In response to the military's increasing use of space-based GIS (Geographic Information Systems) tools, the ISRO has launched the CARTOSAT-1 in 2003 and plans to follow it up with CARTOSAT-2 and RISAT in late 2004–05 and 2006 respectively.

India's indigenous satellite navigation and positioning system 'GAGAN'–GPS and GEO Augmented Navigation is also primarily in response to its Ministry of Civil Aviation and is planned to become a part of the satellite-based CNS/ATM (communication, navigation and surveillance/air traffic management) system for civil aviation. No known military applications or military application proposals to enhance positioning or targeting capabilities exist.

To further augment and enhance its space surveillance capabilities an ambitious SBS (Satellite Based Surveillance) programme was mooted in 1993, but is yet to take off.

With regard to communications, the G-SAT series are known to carry a very limited payload for military use in an overwhelmingly civilian and commercially-driven endeavour.

The dismal state of Indian military space utilisation can be gauged from the fact that though satellite-aided search and rescue is widely available under the COSPAS-SARSAT (Search & Rescue Satellite Aided Tracking) programme and a GEOSAR satellite based system is supported is supported by INSAT-2A, Indian military aircraft continue to use traditional beacons or ELTs (Emergency Locator Terminals) and perforce have to rely on detection by LEO (Low Earth Orbiting) satellites or other traditional methods. Their civilian counterparts are known to have undergone the transformational change quite a while ago.[111]

However, certain institutional efforts appear to be in the offing considering that the Indian Parliamentary Standing Committee on Defence had recommended since the year 2000, the development of an "Aerospace Command". The Indian MoD (Ministry of Defence) in an "action-taken reply" on the recommendation contained in the seventh report on 'modernisation of the Indian Air Force' (Thirteenth *Lok Sabha)* stated that the IAF would undertake preparatory work of defining viable concepts and drafting various doctrinal and command/control models for final approval of the government in regard with formation of an Aerospace Command. This work would largely be exploratory, involving interaction with numerous agencies and academia, thus generating possible options and concepts, and establish the core for such an Aerospace Command. The Standing Committee on Defence reiterated their recommendations in the fourteenth report on action taken and desired that this project should be taken up seriously to ensure that India comes onto the global space map. The MoD was asked to furnish a brief note on the development of the Aerospace Command in the IAF during the examination of demands for grants of the MoD (2003–2004). In response, the MoD stated in a written reply that there was no Aerospace Command in the IAF. Foreign collaborations, if any, were done by the nodal agency—ISRO—in matters relating to aerospace. In response to another question on what extent the development in space technologies could be utilised by the IAF, the MoD stated that developments in space technologies could be utilised by the IAF in the following manner:

- To build real-time situational awareness through space communication and space sensors.
- To link radar and other communication networks over the entire span of the country.
- To assist in Ballistic Missile Defence.
- To gather real-time intelligence about enemy aircraft, missiles, and space-borne threat.
- To prevent the enemy from using own space assets by jamming.

The Standing Committee was unhappy to note that despite its recommendations made in the year 2000, no concrete groundwork has been done so far by the IAF on the development of an Aerospace Command. The Committee emphasised the "phenomenal scope for use of space technologies for Air Defence" including the necessity for protecting the country's assets in space in view of the fact that space warfare involving surveillance monitoring and jamming techniques would be a distinct possibility in the near future. The Committee once again reiterated that the establishment of an Aerospace Command should be taken up seriously by the MoD with planning and groundwork initiated at the earliest.[112] Till date the status quo prevails, no Aerospace Command has been formed and plans if any are yet to take off and are not expected to do so considering that as in September 2005, the present Indian government, like the previous one, turned down the IAF's proposal for an aerospace command.

Indian Castles and Forts in Space

Notwithstanding the above facts, speculation on Indian military space capabilities based on independent assertions by military analysts, academias, etc. or misinterpretation of innocuous statements of military personnel is rife. In fact, the paranoia actually spawns an entire cottage industry for non-proliferation and disarmament activists who attribute fantastic military capabilities to a non-existent programme. Most such misconceptions build upon

- Unsubstantiated assertions by unknown obscure internet sites like *"Bharatvarsha"*.

- Convolution of innocuous statements by military brass like "any country on the fringe of space technology like India has to work towards such a command (aerospace command)o as advanced countries are already moving towards laser weapon platforms in space and killer satellites.[114]

The examples in Figure 6.3 would serve to validate this view.

Speculation on Indian Military Space Capabilities

Author & Institute	Statement	Source of input
Theresa Hitchens, CDI	India has been working on laser technology for military applications since 1990.[245]	Bharatvarsha-1947
Yorkshire CND	India had started development of an aerospace command station to control space-based weapons for a retaliatory capability in case of a nuclear attack by Pakistan.[116]	*Daily Times,* Pakistan
Jeffery Lewis, CDI	Indian military had conducted studies on space based DEW, space-based lasers and Kinetic attack loitering interceptor, etc.[118]	Pronouncements of independent defence analysts

Figure 6.3

The fact that towards the end of 2005, no aerospace command has been formed and that the Indian status quo on space prevails clearly indicates that the above programmes were based on ill-informed speculation and not facts. The facts clearly point to the contrary, far from space weaponry which is clearly way beyond its technological and economical capabilities, the Indian military is yet to put its act together and obtain a reasonable semblance of military space utilisation.

Conclusion

From the foregoing description, it is apparent that India continues to remain oblivious to the military space affairs in its neighbourhood and also the tremendous utility of space for the military. The Chinese Dragon on the one hand surreptitiously pursues a space militarisation and weaponisation programme, and on the other also vociferously protests and calls for preventing an arms race in outer space at international summits.[119] While it complains itself hoarse against the US dominance in space, it is conveniently oblivious to upsetting the 'apple-cart' within Asia. In fact Chinese pronouncements of "thought torments" on account of India's space achievements[120] only serve to lend a more suspicious perception of the obvious disconnect between its words and actions. Meanwhile the great Indian tiger continues snoozing.

Other Aspirants

Apart from the previously covered Asian space powers, other Asian nations ranging from South Korea to Vietnam, Kazakhstan, etc, are also known to harbour grand aspirations and ambitions with regard to space and have space activities at varying degrees and stages. While the usual riders of high technology, cost, and policy issues restrict the acquisition of space capabilities, a lot of Asian countries have not only an expressed interest in both the civil and military applications of space but have also begun an earnest pursuit of space capabilities in a variety of ways. With the presence of commercial space providers proliferating in the open market, it would only be a matter of time before space-enabled civil and military capabilities become readily available. In the absence of clear-cut regulations and legality on the subject being vague, the potential for converting the benign uses of space to more malevolent use is immense and hence a brief perspective would be in order and is undertaken as below.

THE SOUTH KOREAN SPACE PROGRAMME

Early Developments

South Korea's later start in the space race has less to do with disarmament, morality or pacifist notions and is more a consequence of missed opportunities forged out of the special circumstances generated on account of a memorandum of understanding signed with the US which restricted Korea's launch endeavours. A 1979 memorandum with Washington held Seoul's missile range to 180 km, far short of most targets in the North.

Washington imposed and maintained strict controls to avoid sparking an arms race in Northeast Asia. In return for acceding to these terms, Washington offered Seoul missiles and assistance in other defence areas. A direct consequence of this was that Seoul's BM and by extension SLV programme was severely restricted and stunted, severely impeding its growth in space technology and other related areas. Surrounded by economic and strategic competitors like Japan and China with developed commercial and ballistic missile programmes, Seoul lobbied intensely for an end to the restrictions and access to missile technology from other MTCR nations. South Korea sought technological and economic benefits from a purely indigenous space programme, one that could eventually defend the entire peninsula while decreasing economic ando security dependency on the United States.

After negotiating for up to five years, South Korea was finally able to extricate itself from the memorandum and Washington formally agreed to abandon the 1979 agreement in early January 2001, predicating its decision on several factors including the following,

- First, despite restrictions on Seoul, North Korea continued to develop longer-range ballistic missiles, launching a three-stage *Taepo Dong 1* in August 1998. Pyongyang's programme was driven by more than simply a desire to militarily counter the South—to increase its strategic position in relation to allies China and Russia and foes Japan and the United States.

- Second, South Korea argued that the restrictions on its commercial space programme were a serious impediment to economic competitiveness. The agreement did not just limit Seoul's military missiles programme, but its commercial space development as well and restricted access to foreign rocket technology. Seoul contended that to remain competitive in the international aerospace and electronics industries, particularly against neighbouring Japan, South Korean companies needed the technological spin-offs from a domestic space programme.

- Third, and perhaps most important, Washington lifted restrictions to gain access to South Korea's active missile development programme. South Korea's domestic missile and nuclear programmes in the 1970s triggered Washington's initial restrictions. By 1978, South Korea's Agency for Defense

Development (ADD) had already succeeded in converting US-supplied Nike Hercules surface-to-air missiles into ballistic missiles with ranges between 150 and 250 km.[121]

Thus, by January 2001, Washington and Seoul formally abandoned the 1979 memorandum and Seoul announced its intentions to join the Missile Technology Control Regime (MTCR) which would allow it access to rocket technology from other MTCR members. Ever since, South Korea has pursued a highly indigenised programme premised upon the following objectives:[122]

- Capability to launch micro satellites indigenously by the year 2005.
- Indigenous development of low earth orbit multi-purpose satellite and launcher by the year 2010.
- Plan to enter into worldwide top 10 countries in the space industry by the year 2015.

All space development in South Korea is premised upon its National Space Development Plan issued in 1995 until 2015. This space development plan is composed of three areas: rocket development, satellite development, and space science and application. The final goal is for Korea to become one of the top ten space-faring countries by 2015.

The National Mid-Long Term Space Development Plan of Korea envisions the launching of 20 satellites by the year 2015, including five communication satellites, eight multi-purpose satellites, and seven scientific satellites. The objective of this plan, whose budget is estimated at US$4.8 billion is to establish an independent domestic satellite technology capacity by the year 2015. It includes development of technology and human resources in the fields of communication satellites, Earth observation satellites, and launch vehicles for low earth orbit satellites.[123]

Organisation and Infrastructure

As in the case of most nations, Korea develops space as an intrinsic element of its composite aerospace power and this is reflected in its organisational framework. In Korea, air and space development is divided into three parts: space development, air development, and air and space defence system. Space development in Korea is driven by the view of national needs and a technology development strategy. As a result, the

Ministry of Science and Technology takes the leading role in the space development, while air development is mainly the responsibility of the Ministry of Industry in terms of industrialisation and commercialisation. However, the development of air and space defence system is the sole and direct responsibility of the Ministry of Defence. The Agency for Defense Development of Korea (ADD) takes on the actual responsibilities in the weapon system developments. However, except for weapon systems, the Korea Aerospace Research Institute (KARI) plays the leading role in all air and space development activities in Korea. KARI is a government-funded research institute established in 1989 based on a special law to promote the development of air and space.[124]

Launch Vehicles

Washington's 1979 memorandum aimed at stemming missile development in North-East Asia had contrarily prompted South Korea to carry out a clandestine programme to match its neighbours like North Korea, China, Japan, etc. In fact, in the late 1970s (1978) South Korea's Agency for Defence Development (ADD) was known to have succeeded in converting US-supplied Nike Hercules surface-to-air missiles into ballistic missiles with ranges between 150 and 250 km and even later in the 1980s and early 1990s, it continued with its clandestine efforts. By 1993, it had succeeded in launching its single-stage sounding rockets and in 1997 and 1998; the two-stage sounding rocket KSR-2 was also successfully launched for space science missions.

On 28 November 2002 KARI succeeded in launching its first single-stage liquid fuel rocket KSR-3e from a launch pad in *Anheung* in the central province of South *Chungcheong*. The development of KSR-3 with its liquid propellant engine will be the basis for Korea Space Launch Vehicle 1 (KSLV-1). KSLV-1 is designed to put small (100-kg-class) satellites into LEO orbit. It is scheduled for launch in 2005. Korea's long-term objectives as per its national plan are:

* Acquiring the capability for the indigenous development of a low Earth orbit satellite launcher.
— Indigenous launch of low Earth orbit micro satellites in 2005.
— Indigenous launch of low Earth orbit multi-purpose satellites in 2010.
● Construction and operation of a space centre by 2005.

Satellite Programmes

It terms of satellite possessions, South Korea ranks eighteenth at the global level and its satellite development primarily comprises the following three programmes;

- Experimental satellites, KITSAT (*Uri-Byul*).
- Commercial satellites, KoreaSat *(Mugunghwa)*.
- Multi-purpose Earth observation KOMPSAT (Korea Multi-Purpose satellites).

KITSAT

South Korea joined the space race with the launch of its first 100 kg micro satellite (microsat) *KITSAT-1* on 10 August 1992. This was followed by *KITSAT-2* on 26 September 1993. Both the satellites were based on UK's Surrey Satellite Technology Limited (SSTL) microsat design and were carried as piggy back passengers on Ariane flights to LEO. Both of these were scientific satellites, provided with a modest store-and-forwarding messaging capability for communications also. Its third microsat in the above series, *KITSAT-3* was launched successfully on 26 May 1999 and continues to be in operation. *KITSAT-4* is under development and likely to be launched soon.

KoreaSat

The KoreaSat series are commercial broadcasting and communication satellites of the Korean Telecommunication (KT) Corporation. *KoreaSat-1* (1995), *KoreaSat-2* (1996) and *KoreaSat-3* (1999) were developed and successfully launched through overseas contracts. *KoreaSat-3* is a 110 kg small satellite with three axis stabilisation capability to support a 3-band CCD (Charge Coupled Device) camera at a circular orbit of 720 km altitude. *KoreaSat-3's* military communication transponders are known to provide service for military communications.[125]

On 9 January 2002, the South Korean Ministry of Information and Communication (MoI&C) announced that South Korea would launch its first military satellite in 2005. The primary investors in the US$ 410 million project are the MoD financing $180 million and the MoI&C and local communication giant KT funding the remaining $230 million.

KoreaSat-5 would be the fourth satellite launched by Korea, replacing *KoreaSat-2*, whose lifespan is due to expire by 2005. Though it is to be a dual-purpose satellite, with both military and commercial uses, *KoreaSat-5* would be the first Korean satellite with a military mission. The satellite will be placed at an altitude of 38,000 km from the equator and allow the military to conduct operations without the hindrances associated with wiretapping or radiowave warfare. It will also help the nation in developing a variety of satellite-guided munitions. The new satellite was named *KoreaSat-5* instead of four, because the Sino–Korean word for the number four, "sa" is a homonym of the Chinese character for death, according to the Korean government.[126] *KoreaSat-5* is expected to be shared equally by both private users and military authorities and shall be managed by Korean scientists upon its launch. Presently, "preparation delays" have led to postponement of its launch till 2006.

KOMPSAT

KOMPSAT is a key programme of Korea's space development plan and is intended to be the backbone of Korea's satellite technology development. KOMPSAT-1's development started in 1994 and the satellite was finally launched in December 1999. KOMPSAT-1 is an Earth observation satellite and has three payloads composed of EOC, OSMI, and SPS, as well an 8-Gb solid-state data storage unit. EOC is a high-resolution panchromatic camera of 6.6-m GSD and 17-km swatch. OSMI is the ocean-scanning multispectral camera of 1-km GSD and 800-km swath. The six bands of OSMI can be selected by ground control using its internal hyperspherical capabilities. SPS is the science physics sensor for ion measurement and high-energy particle detection at the KOMPSAT orbit.

KOMPSAT-2 is presently under development and is planned for launch in 2004. The main payload of KOMPSAT-2 is its multispectral camera which features a 1-m GSD panchromatic band and 4-m GSD 4-colour band with 15-km swathe.[127]

In addition to the above, Korea has spelt out its long-term objectives with respect to satellite development as below:

- Development and launch of a total of 20 satellites by the year 2015.

- Building up the capability to develop LEO multi-purpose satellites indigenously.

- Acquiring the capability of satellite data processing and its application technology.

Developmental Strategy

To make up for lost time, Korea has adopted a mid-entry strategic approach which would enable it to leap-frog to requisite technology levels through technology transfers. The joint development of KOMPSAT with the US is a typical example of the same effort. In addition, Korea has also adopted a strategy of smart selection, whereby it selects areas of development based upon its unique needs and resources instead of trying to accomplish overall proficiency in all areas. For space development, Korea endeavours to develop a 500–1,000-kg class satellite and its launching capability to LEO orbit. For space science and applications, Korea just wants to keep up with the leading countries through international cooperation.

THE INDONESIAN SPACE PROGRAMME

The Indonesian space programme began in the mid-1970s and is characterised primarily by a desire to link its dispersed islets and mitigate the challenges posed to its integrity and promote national unity by enhancing communicability within the nation and overseas. Indonesia's status as an archipelagic state with about 17,000 islands, more than 200 million people and up to 200 different languages poses enormous problems on connectivity and communications which Indonesia has sought to allay using communication satellites. Due to its island geography, Indonesia sought to develop an extensive satellite communications system, and on 17 August 1976 in an attempt to electronically unify the nation, *Palapa-A1* built by America's Boeing Satellite Systems was launched for Indonesia by NASA on a Delta-2914 launcher from Kennedy space centre, Florida. As an interesting aside, *'Palapa'*, a name signifying national unity, was chosen by President Suharto in July 1975. The name symbolises the fulfillment of a vow for unity first expressed by GajahMada a revered national hero of the 14th century who served as Prime Minister of the Kingdom of *Majapahit*. He had vowed not to partake of *Palapa*, a national delicacy, until the goal of national unity was achieved.[128]

Palapa

The Palapa-A satellite programme began in February 1975 when the Indonesian government awarded two separate contracts to Boeing for the construction of two satellites, a master control station for the entire system, and nine earth stations. Completion of the earth stations and development, construction, and launch of the first satellite took place in 17 months, other firms built 30 earth stations to complete the group segment of the system, which was controlled and operated by PERUMTEL, a government-owned telecommunications company. *Palapa-A*'s specially designed antenna concentrated the satellite's signal power on all Indonesian islands, including the main isles of Sumatra, Java, Bali, Kalimantan, Sulawesi, and Irian Jaya, as well as the surrounding southeast Asia area including Singapore, Malaysia, Thailand, and the Philippines. This was followed by the *Palapa-A2* on 10 March 1977; both the spacecraft were identical to Canada's Anik and Western Union's Wasters except for a modified parabolic reflector, enlarged to give maximum illumination of the Indonesian land mass. Operational lives for *Palapa A1* and *A2* ended on June 1985 and January 1988, respectively.[129] The Palapa B series (double in size and capacity) continued in the 1980s and 1990s providing communication links to Indonesia as well as New Guinea, Malaysia, Philippines, Singapore and Thailand. Privatised in 1993, the Palapa satellites business is now operated by PT Satelindo, which is controlled by Bambang Trihatmodjo, a son of President Suharto. *Palapa-B* was the second generation of communications satellites designed and built for Indonesia by Hughes Space and Communications Company. This series of four satellites were produced for Indonesia's state-owned telecommunications company, Telkom. All the Palapa-B spacecraft are Hughes HS-376 models. These were followed by the Palapa-C series which were version of the HS-601 in the mid-1990s with the first of the series being launched on 31 January 1996, on an Atlas-2AS booster. All of these are depicted in Figure 6.4. Each succeeding generation of Palapa satellites was significantly larger and more powerful than the one it replaced, as demand for services grew and not only was Palapa-C larger and more powerful, it also covered southeast Asia and parts of China, India, Japan, and even Australia.

Palapa Satellite Launches

Satellite	Date	Launcher	Remarks
Palapa A1	08-07-1976	Delta 2914	
Palapa A2	10-03-1977	Delta 2914	
Palapa B1	18-06-1983	Shuttle, PAM-D	With Challenger F2, AnikC2, SPAS.
Palapa B2	03-02-1984	Shuttle, PAM-D	With Challenger F4, Westar6, IRT
Palapa B2P	20-03-1987	Delta 3920, PAM-D	Ex Palapa B3
Palapa B2R	13-04-1990	Delta 6295-8	NewSat-1
Palapa B4	13-05-1992	Delta 7925-8	
Palapa C1	01-02-1996	Atlas-2AS	Later Anatolia-1 and now Paksat-1
Palapa C2	16-05-1996	Ariane-44L H10-3	With AMOS-1

Figure 6.4

This was followed by IndoStar/Cakrawarka-1 for television broadcasting in 1997 until the recent TELEKOM series for fixed communication and broadcasting and *Garuda-1* satellite for personal global mobile communication.

Indonesia is set to launch its first imagery viewer micro-satellite in 2005 to carry out remote sensing of the territory and its natural resources. The 50-kg Lapan-Tubsat satellite will be launched by an Indian PSLV and will ride piggyback with an Indian satellite on the rocket during the last quarter of 2005 or early 2006. The micro-satellite is a result of collaboration between Indonesia and Germany's Technischen Universitat Berlin and will be completed in 18 months at a cost of US$2 million. "The satellite will prevent the Bawean incident from recurring", Air Force Chief Marshall Chappy Hakim was quoted by *The Jakarta Post* as saying. He referred to the five American F-18 Hornet jets that trespassed into Indonesian airspace in July 2003 and performed manoeuvres for more two hours over Bawean Island in the Java Sea. The United States claimed that it had secured permission to enter Indonesian airspace while escorting a US aircraft carrier, two frigates and a tanker. The government, however, negated the claim, saying the request had arrived too late at air defence command. The case could be just one of the many undetected breaches of Indonesia's territorial integrity due to the vastness of the archipelagic country. Thus far, Indonesia has been paying $250,000 a year for the use of imagery viewer satellites of other countries.[130]

Space Organisation and Infrastructure

In order to promote space in a comprehensive and systematic manner, Indonesia has created a number of agencies, firmly networked, both commercial and governmental which have been briefly described as below:

Depanri

The highest coordinating body in formulating aerospace policy in Indonesia is the National Council for Aeronautic and Space (DEPANRI). DEPANRI was established by Presidential Decree no. 24 of 1963 as amended by Presidential Decree no. 99 of 1993. DEPANRI is chaired by the President of the Republic of Indonesia with members comprising the Minister of State Research and Technology, Minister of Foreign Affairs; Minister of Trade and Industry; Minister of Communication; Minister of Defence; and Minister of State Development Planning.

The main duty of DEPANRI is to assist the President in formulating general policy in the field of aviation and space. DEPANRI is obliged to provide consideration, opinion and advice to the President regarding legislation and utilisation of air space and outer space for aviation, telecommunication and other national interests.[131]

The National Institute of Aeronautics and Space (LAPAN)

LAPAN acts as a national focal point in conducting research and development related to the peaceful uses of outer space. LAPAN is directly responsible to the President of Indonesia while its activities are technically coordinated by the Ministry of State for Research and Technology. Its main functions include the utilisation of remote sensing satellite data and undertaking activities related to research and observations of the atmosphere/upper atmosphere.[132]

Other Governmental Institutions

Other governmental institutions involved in space technology applications are: The National Coordinating Agency for Surveying and Mapping (BAKOSURTANAL); The Meteorological and Geophysical Agency (BMG); The Agency for the Assessment and Application of Technology (BPPT); The Indonesian Institute of Sciences (LIPI).

Non Governmental Institutions

The primary non governmental institutions involved in space-based activities in a big way are the Indonesia Satellite Association (ASSI) and Indonesia Infocom Society (MASTEL).

THE MALAYSIAN SPACE PROGRAMME

Malaysian space endeavours are a fairly recent development. On 12 January 1996 Malaysia became a space-faring nation when it launched its telecommunications satellites, MEASAT-1 (Malaysia-East Asia Satellite-1) aboard an Ariane from Kourou, French Guiana. This was followed by *MEASAT-2* on 13 November 1996, also aboard an Ariane. Both MEASAT spacecraft were built by Hughes Space and Communications Company (HSC) in El Segundo, California, now also Boeing Satellite Systems. In March 2003, Binariang Satellite Systems Sdn. Bhd. of Malaysia ordered a high-power commercial communications satellite from Boeing Satellite Systems, a new Boeing 601HP satellite, designed *MEASAT-3,* to join the existing Boeing-built *MEASAT-1* and *MEASAT-2* spacecraft. *MEASAT-3* is slated for launch in 2005 aboard a Proton Breeze-M from Baikonour Kazakhstan.

Malaysia has also built its first micro satellite, *TiungSAT-1,* in collaboration with the United Kingdom. Named after a variety of a singing mynah bird, the satellite which was launched on 26 September 2000, operates on amateur radio frequencies and has remote-sensing capability. In view of the great potential for applications and affordability of small satellites, Malaysia is committed to the research and development of such satellites and to exploit its advantages in new ways. A second micro satellite (RazakSat) designed for low Earth orbit at the equator is under way. This satellite will be able to cover 51 nations, most of them developing and located near the equator. The satellite is slated for launch in 2005. The Malaysian Prime Minister Datuk Seri Dr Mahathir Mohamad had also announced in 2001 that Malaysia would soon launch eight small satellites in orbit under its "NeqO" space programme.[133]

Malaysia's National Space Agency, which is under the Ministry of Science, Technology and environment is responsible for coordinating Malaysia's requirements in aerospace and satellite technology. It has also been entrusted with the tasks of identifying the necessary infrastructure, formulating national space policies, and planning space and satellite programmes.

THE IRANIAN SPACE PROGRAMME

Iran in its pursuit of space-based capabilities has declared that "Special emphasis has been put on the practical uses of space science and technology in the areas of remote sensing, satellite communications, broadcasting, meteorology and satellite navigation". In order to coordinate all ongoing activities in research-oriented institutions, administrative agencies, and universities, a recently modified national body called the "Iranian National Committee on Peaceful Uses of Outer Space (INCOPUOS)" has been formed to act as the focal point responsible for policy and decision making. These efforts arise from Iran's conviction that in spite of high cost of involvement in space activities, it needs to take advantage of space's practical benefits for economic and social developments.[134]

Satellite Programmes

While Iran had sought Communications satellites since 1980, and had also reportedly finalised a deal for the manufacture of its first *"Zohreh"* (Venus) satellite in March 2001 (as part of its domestic satellite communications programme), not much was known about its space ambitions in open literature, and its space ambitions and programmes, if any were nascent up to the last decade. The first official indications of its ambitions were apparent at a Teheran Aerospace Conference ("Space and National Security"), in January 2004, when its defence minister Ali Shamkhani vowed that Iran would launch a satellite of its own within 18 months. The minister's pronouncement that "Iran will be the first Islamic country to enter the stratosphere with its own satellite and its own indigenous launch system," is generally considered to be the first official indication of a time frame for Iran's space programme. Mr Shamkhani described Iran's aerospace programme as being part of the country's "deterrence force", and said universities and the defence industry were cooperating on the satellite, but gave no details as to what type of satellite was planned.[135] By September 2004, Iranian media reported that Iran intended to launch its first satellite into space in April 2005, with the device described as being purely for civil purposes. The satellite, codenamed *"Mesbah"* (lantern), was shown on state television. It was said to weigh 60 kg (132 pounds) was cube-shaped with each side measuring 50 centimetres (20 inches).[136] It would be put into orbit at an altitude of 900 kilometres (about 560 miles). However, the Mesbah project is apparently behind schedule considering that as early as 27 June 1998 an agreement in Iran was signed to design, construct and launch a communications research, Earth-observation satellite *Mesbah,* and a small technological satellite. This agreement was with the Minister of Higher Education, Mostafa Moin and the Minister of Post, Telegraph, and Telephone, Minister Reza Arefvazdi. The *Mesbah* satellite was to be launched within the next three years (2001) of signing with apparently China and Russia's Glavkosmos organisation as a piggy-back payload on Russian and Chinese boosters. The project was to be run by The Scientific and Industrial Research organisation of Iran and its Center for Telecommunication Research. This is one of several communication satellite projects being developed by Iran.

Besides the communication satellites, two mini-satellites are being developed. The (Small Multi-Mission Satellite) SMMS is a joint venture payload between China, Iran, South Korea, Mongolia, Pakistan, Thailand and Bangladesh under the Asian-Pacific organisation. This SMMS imaging spacecraft payload project is primarily managed by China and Iran. It was to be launched on a Long 4 March in 2000 and 2001 but has subsequently been delayed to 2005 as a part of the China's weather satellite programme. The launch will place the SMMS spacecraft into a 650-km Sun-synchronous polar orbit. The 470-kg satellite has increased it mass from its original 380-kg and is based on the CAST-968B platform developed by China's space industry of the PRC Academy of Space Technology. It will carry a 100-kg multi-spectral CCD imaging camera. The camera is capable of wide field 20 metres resolution imaging.[137] On 3 October 2004, Deputy Minister of Information and Communication Technology (ICT) and Head of Iran's Space Agency Seyyed Hassan Shafti stated that the multipurpose small satellite costing $44 million is scheduled for launch in 2006.[138]

Launch Vehicle Programmes

As regards indigenous launch capabilities, Iran is known to be committed to the development of the space booster "IRIS". Analysts on the subject speculate that the IRIS launch vehicle apparently consists of the No-Dong/Shahab-3 first stage with a bulbous front section ultimately designed to carry an additional second stage solid motor as well as a communications satellite or scientific payload. The IRIS launch vehicle is a space related derivation of the *"Shahab-3" (Meteor-3)* ballistic missile. A launch vehicle of this configuration is ideal as a vertical probe sounding rocket for ballistic warhead re-entry vehicle development or a scientific payload; however, it would almost certainly not be capable of launching a satellite of appreciable mass or capability unless it were intended to be a second and third stage of a larger launch vehicle. In 1998, Iran had also announced plans to build a telecommunications satellites, to be launched in 2001 using a Shahab-4 rocket. Iranian Defence Minister, Admiral Ali Shamkhani had announced that Shahab-4 is now in production as a space launcher. "Shahab-4 is not for military purposes but for launching a satellite".[139] Nevertheless, amidst Iran's governmental pronouncements that it was deliberately being ambiguous about its launch capabilities, the country has announced it has upgraded the *Shahab-3,* and has denied it is working on a Shahab-4, a device that would involve a two-stage propulsion system and bring European capitals within range. However, Iranian capabilities began to appear more credible subsequent to its successful test of an upgraded version of its 2000-km range *Shahab-3* missile on 11 August 2004. Former president Akbar Hashemi Rafsanjani subsequent to the launch stated that "with this ballistic power, we can today speak of an independent satellite launch and we should seek the technology to make our own satellites". Thus, it can be safely inferred that by late 2004 or early 2005, in view of Iran's increasing proficiency in missile technologies, SLV technology would also become available to Iran and in most likelihood, its space efforts would cross the stratosphere and be in orbit.

NORTH-KOREAN SPACE ENDEAVOURS

On 04 September 1998 the Korean Central News Agency broadcast a report claiming the successful launch of the first North Korean artificial satellite, *Kwangmyongsong-1"* (Bright Star-1). North Korea tried to launch into orbit a "very small satellite" on 31 August 1998, according to that nation's government as well as the US State Department and the South Korean Foreign Ministry. Apparently the satellite failed to reach orbit as it could not be found in space by US military and other trackers. As of 09 September 1998 US Space Command had not been able to confirm

North Korean assertions. US Space Command had not observed any object orbiting the Earth that correlated to the orbital data the North Koreans had provided in their public statements. Initial reports that Russian military space forces had confirmed that the satellite was in orbit had subsequently been withdrawn.[140] As of now, the speculation has been laid to rest and most literature on the subject confirm the view that the satellite was a North Korean launch of its first medium-range *Taepo Dong 1* ballistic missile from the north eastern part of North Korea shortly after noon on 31 August 1998. The rocket landed in the high seas off the Sanriku coast of Japan, after flying over the Japanese island of Honshu before plunging into the Pacific Ocean, trigerring off Japan's and South Korean space militarisation programmes.

Conclusion

In addition to the above-mentioned countries, a fairly large number of Asian nations ranging from the Arab consortium, Thailand to Kazakhstan, Vietnam, etc. aspires to have capabilities in space and known to be expediting efforts to attain their goal for reasons unique to individual national needs and requirements. For example, while Arab nations like Saudi-Arabia, UAE, etc. procure space capabilities in the open market as per their requirements, countries like Kazakhstan, by virtue of being cosmodrome locations of the FSU (Former Soviet Union) are trying to harness space capabilities available by default for promoting national interests. In fact, Kazakhstan also plans to put its own communication satellite in orbit by 2005, paying Russia about $65 million to design and launch its communication satellite. Thus by the next few years the number of nations pursuing space-based capabilities in Asia may not only rival, but may also exceed those in the Western hemisphere and considering that the military option of space is also being pursued vigorously, it would only be a matter of time before the paradigm of space is added to the present security calculus of nuclear and missile developments.

Notes and References

1. For a more detailed brief covering both civilian and military aspects of the Chinese Space programme refer to Sqn Ldr KK Nair, "China's Space Programme: An Overview" in *Airpower Journal*, vol. 1, no. 1, Monsoon 2004, pp. 125–156.

2. The tremendous impact of space-based RMA in promoting Information

dominance and the seminal role of Information dominance in influencing the outcome of wars is well comprehended by the Chinese and hence the accent on "speeding up informationalisation". For further details on informationalisation refer Chinese govt. white paper, *China's National Defence in 2004,* Chapter 3 at www.china.org.cn/e-white/20041227/III.htm

3. For a more complete brief on the doctrinal drivers of the Chinese space programme, see Sqn Ldr KK Nair, "Space Theories and Doctrines: A Comparative Overview" in *Airpower Journal,* vol. 1, no. 3.

4. See William S. Murray and Robert Antonellis, "China's Space Programme", *Orbis,* Fall 2003.

5. US Department of Defence (DoD) FY-04, "Annual Report on the Military Power of the People's Republic of China", pp. 47.

6. Estimated numbers contained in Indian military publications like SP's and Indian Defence Year Book for the year 2005 are identical to Mil Balance (2003-04) and hence are not mentioned separately.

7. *Jane's Intelligence Review,* December 2003.

8. Ibid.

9. John Pike, "The military uses of outer space", *SIPRI Year book 2002: Armaments, Disarmaments and International Security.*

10. "Dragons in Orbit? Analysing the Chinese approach to space", National Defence University, Washington DC available on www.ndu.edn/inss/china_center/paper10 htm

11. US DoD FY-04 report, p. 41.

12. Zhu Yilin, "Fast track Development of Space Technology in China", *Space Policy,* May 1996, p. 139.

13. Scott Pace et al., "The Global Positioning System: Assessing National Policies", Santa Monica: Calif.Rand.Critical Technologies Institute, MR-614-OSTP, 1995, p. 68.

14. See William S. Murray, *Orbis,* Fall 2003.

15. Paul Beaver, *Jane's Defence Weekly,* 02 December 1998, p. 18.

16. The US Department of Defence, "Annual Report on the Military Power of the People's Republic of China" July 2003, p. 36 also claimed that China was developing killer microsatellites based largely on a January 2001 Hong Kong newspaper article. However, Gregory Kulacki and David Wright of the Union of Concerned Scientists traced the story to a website run by a self-described 'military enthusiast' named Hang Chaofei who ran a Chinese language Internet bulletin board filled with crude illustrations and "fanciful stories about secret Chinese weapons to be used against Americans in a future war over Taiwan". For details, see Gregory Kulacki and David Wright, "A Military Intelligence Failure? The Case of the Parasitic Satellite, Cambridge, MA: Union of Concerned Scientists, August 2004, p. 3.

17. Matthew Mowthorpe, *"The Militarisation and Weaponisation of Space"*, Lexington Books 2004, p. 102.

18. Cheng Ho, *Space Daily,* 08 July 2000.

19. The above estimate is attributed to Air Commodore Jasjit Singh. See Air Commodore Jasjit Singh, "Force Modernisation Planning Challenge", *Airpower Journal,* Vol. 1, No. 2, Oct–Dec 2004, p. 26.

20. Williams S. Murray, *Orbis,* Fall 2003.

21. During January 1996, the United Nations (UN) International Telecommunications Union (ITU) supported the Pacific Telecommunications Conference to address both GEO crowding and frequency allocations and developed a number of suggestions to alleviate these problems. Only a few months later, as reported by the UN themselves, severe crowding in the geostationary orbital slots over Asia led to the jamming of a communication satellite by PT Pasifik Satellite Nusantara (PSN) of Jakarta, Indonesia, in defence of an orbital position claimed by Indonesia. The government of Naura, over whose territory the satellites was providing services, charged Indonesia with jamming the satellite. This incident focused global attention on a worsening problem of orbital crowding and caused the matter to be brought before the October–November 1997 World Radio communication Conference (WRC) of the 187 member-nation ITU in Geneva. The conference, after nearly six weeks, made only minor modifications to the procedures for reserving orbital slots and came to no resolution as to the Indonesian jamming incident. The ITU did not get involved in the dispute settlement process, claiming that bilateral negotiations were appropriate.

22. For Chinese recommendations on space-based infrastructure, refer Sqn Ldr KK Nair, "China's Space Programme", *Airpower Journal,* p. 155.

23. See "Pakistan derives its first Hatf missile from foreign space rockets", *The Risk Report,* Vol. 1, No. 8, p. 4, 11 October 2004.

24. Anthony R. Curtis, "Space and beyond: Pakistan's new moon", American Radio Relay League, Inc, 25 July 2003 from www2.arrl.org/news/features/2003/07/10/1.htm

25. Muhammad Irshad, "Pakistan's Satellite", *Pakistan's Defence Journal* from www.defencejournal.com/2002/nov/pak-satellite.

26. FAS on SUPARCO at www.fas.org/spp/guide/pakistan/agency/

27. Muhammad Irshad, "Pakistan's satellite" *Pakistan's Defence Journal* from www.defencejournal.com/2002/nov/pak-satellite.

28. "Pakistan derives its first Hatf missiles from foreign space rockets", *The Risk Report,* Vol. 1, No. 8, 11 October 2004.

29. NTI "Country overviews: Pakistan: Missile facilities", from www.nti.org/e_research/profiles/pakistan/missile/3294_3329.html

30. CNS-space: Pakistan: Launch capabilities from cns.miis.edu/research/space/Pakistan/launch.htm

31. "Pakistan's Badr-B satellite to go into orbit next month", *The News,* Islamabad, March 1, 2001.

32. Refer Dr Shireen M Mazari's paper "Nature of Future War in South Asia" presented in seminar on Aerospace Power in South Asia at Centre for Aerospace Power Studies, Karachi, Pakistan available at www.caps.org.pk/Papers/September2002.htm

33. Ibid., p. 8.

34. Anthony R Curtis, "Space and beyond: Pakistan's new moon".

35. FAS on *Badr-B* from www.fas.org/spp/guide/pakistan/earth

36. Quoted from Muhammad Rafique, "Pakistan airs its new force", *Asia Times on line* from www.atimes.com/ind-pak/C107Df01.html

37. CNS-space: Pakistan military programmes from cns.miis.edu/research/space/Pakistan/mil.htm

38. FAS on *Badr-B*....

39. CNS-space: Pakistan military programmes from cns.miis.edu/research/space/Pakistan/launch.htm

40. Ahmed Abbas, "Indian ambitions for aerospace supremacy: options for Pakistan", Institute of Strategic Studies, Islamabad, Pakistan from www.issi.org.pk/strategic_studies_htm/2003/no_1/article/8a.htm

41. Muhammad Irshad, "Pakistan's satellite", *Pakistan Defence Journal* from www.defencejournal.com/2002/nov/pak-satellite.

42. Nadeem Iqbal, "Pakistan scrambles to launch satellite, eyes bigger plans", IPS 02 August, 2002 available at www.spacedaily/com.news/nuclear-india-pakistan-02za.html

43. CNS-space: Pakistan: Military programmes.

44. "Pakistan seeks technical aid for planned satellite purchase", Space news business report, 29 October 2003 at www.space.com/spacenews/archive03/genevabriefsearch_102903.html

45. Apart from seizing control of the ultimate "high ground" during conflicts, Chinese space scientists in 2002 were also known to have urged their government to stake territorial claims in outer space and develop "space territory" as part of Chinese national strategy. For details refer to Sqn Ldr KK Nair, "China's Space Programme: An Overview", *Airpower Journal,* Vol. 1, No. 1, Monsoon 2004, pp. 146, 155.

46. The above was stated by the PAF chief during his inaugural address to participants of a seminar on "Aerospace Power in South Asia" hosted jointly by the Centre for Aerospace Studies, Karachi and the Institute of Strategic

Studies, Islamabad in September 2002. While the requirement of defence of national frontiers in space has already been conceived by the PAF chief, the elemental geographical characteristics required to carve out frontiers in space are yet to be conceived in any credible manner and most such conceptions of geographical distances and territories in terms of light years, etc. are yet to move beyond the realm of fantasy.

47. Refer to statements and replies to queries by Nasim M Shah, Secretary SUPARCO subsequent to his paper "Military Applications of Space Technology: Capabilities and Future Potential of Pakistan" presented at the aforementioned seminar, pp. 25, 31, 32.

48. Gerald Steinberg, "Dual use aspects of commercial high-resolution imaging satellites", *New Besa Publications,* Begin Sadat Center for Strategic Studies, Chapter 4, and the Japanese Constitution available at http://www.solon.org/Constitutions/Japan/English/english-Constitution.html and various other sites.

49. Paul Kallender, "Japan seeks dual-use space technology ok", *Defense News,* 19 July 2004.

50. Gerald Steinberg, "Dual use aspects of commercial high-resolution imaging satellites", *New Besa publications,* Begin Sadat center for strategic studies Chapter 4.

51. FAS on Information Gathering Satellites at www.fas.org/spp/guide/Japan/military/imint

52. Kyle T. Umezu "Early bird tweaks the law", *Japan's space net* at www.spacer.com/spacenet/text/spy-97a.html

53. Axel Berkofsky, "Look up, Mr. Kim: Japan's spy in the sky", *Asia Times,* 15 January 2003.

54. *Nature Asia,* "Japan split over US aid for spy-sats", Vol. 396, No. 6710, 3 December 1998.

55. Motohiro Tsuchiya, "Snooper satellite above East Asia", Center for Global Communications, International University of Japan at ifrm.glocom.ac.jp/gii/taiyo19990607en.

56. United Press International, "UPI hears", Washington DC, 26 August at www.washtimes.com/upi-breaking/20040826-022808-6115r.htm.

57. FAS on Information Gathering Satellites.

58. Ibid.

59. Interim report on the constitution of Japan available at the Research Commission on the Constitution's page on the japanese House of Representatives website.

60. Paul Kallender, "Japan aims for operational military space systems by 2006", *Space News,* 02 September 2003.

61. United Press International, "UPI hears", Washington DC, August 26 at www.washtimes.com/upi-breaking/20040826-022808-6115r.htm.

62. Paul Kallender, "Japan aims for operational military space systems by 2006",

63. United Press International, "UPI hears", Washington DC, 26 August at www.washtimes.com/upi-breaking/2004/2440826-022808-6115r.htm.

64. Paul Kallender, "Japan aims for operational military space systems by 2006".

65. Ibid.

66. *Daily Times*, "Japan to deploy unmanned fighters, mulls US satellite-linked air breathing system", 21 September 2003, p. 4 at www.dailytimes.com.pk/default.asp?

67. ATIP, Chapter 8.

68. Michael Swaine, Rachel Swanger, Takashi Kawakami, "Japan and ballistic missile defence", Rand Publications, *MR-1374-CAPP*, 2001, Chapter 2, pp. 17–18.

69. Ken Jimbo, "A Japanese perspective on missile defence and strategic coordination", *The NON-Proliferation Review*, Summer 2002, Vol. 9, No. 2, Center for Asia-Pacific Studies, Japan Institute of International Affairs.

70. Hiromuchi Umebayashi, "Does Japan need missile defence or missile control", *INESAP bulletin-19 missile defence and north-east Asia*,

71. Nuclear Policy Research Institute, "Japan to deploy own star wars", *Sunday Mail*, 22 June 2003.

72. *Kyodo News*, "Debate over Japan introducing missile defence heating up", 23 April 2003.

73. Gerald M. Steinberg, "Middle East space race gathers pace", *International Defence Review*, Vol. 28, October 1995.

74. Gerald M. Steinberg, "Commercial observation satellites in the Middle East and the Persian Gulf", *Rand Publications*, (MR-1229, 2001), Chapter 11, p. 2.

75. Amnon Barzilai, "Somewhere beyond the horizon", *Harretz.com.news.updates*, 14 October 2004.

76. Ibid.

77. Ibid.

78. NTI, "Israel profile-missile chronology", from www.nti.org/e_research/profiles/Israel

79. "Israel: how far can its missiles fly?", *The Risk Report*, Vol. 1, No. 5, p. 1.

80. "Israel's triad could deter TBM attacks", *Jane's Missile and Rockets*, June 2001.

81. Center for Non-proliferation Studies (CNS), "Israel: launch capabilities" from cns.miis.edu/research/space/Israel/launch.htm

82. "Shavit" from www.israeli-weapons.com

83. David A. Fulghum, "Micro and nano Sats", *Aviation Week & Space Technology,* 16 June 2003, p. 142.

84. CNS, "Israel: Military Programs", from cns.miis.edu/research/space/Israel/mil.htm

85. The first three are Japan, China and India in Asia.

86. Gerald M. Steinberg, "Middle East space race gathers pace", *International Defence Review,* Vol. 28, October 1995.

87. Michael Matza, "Spy satellite aiding Israelis", *Mercury News,* 20 March, 2003.

88. "Israel's Ofeq-5 capable of rapid orbital detuning", *DEBKA File's Military Sources Report,* 28 May 2002.

89. Barbara Opall-Rome, "Israel firm seeks approval for new spy satellite", *Defensenews.com* 22 July 2002.

90. Michael Matza, "Spy satellite aiding Israelis", *Mercury News,* 20 March 2003.

91. "Loss of Ofeq-6 in critical period", *DEBKA File's Military Report,* 7 September 2004.

92. Arieh O' Sullivan, "MoD to investigate *Ofeq-6* launch failure", online edition, *Jerusalem Post,* 6 September 2004.

93. Barbara Opall Rome, "Israel makes plans for broad space capabilities", *Space News,* 25 August 2003.

94. Ibid.

95. CNS, "Israel: Military Programs", from cns.miis.edu/research/space/Israel/mil.htm

96. Gerald M. Steinberg, "Satellite capabilities of emerging space competent states" from faculty.biu.ac.il/steing/military/sat.htm

97. AMOS-1 from www.Israel.weapons.com

98. Ed Blanche, "Israel seeks new high ground to develop spy satellites". *Daily Star* quoted in *Al Jazeerah News,* August 2003 from www.aljazeerah.info.

99. CNS, "Israel: Military Programs".

100. David A. Fulghum, "Micro and nano Sats", *Aviation Week & Space Technology,* 16 June 2003, p. 141.

101. Ibid., p. 140.

102. CNS, "Israel: Military Programs".

103. Jarnal Al-Din Husayn, "Israel: Peace and Arms", *Cairo Rose Al Yusuf,* 21–27 August 1999, (FBIS document ID: FTS19990826001012).

104. Bodhidharma from Word IQ Dictionary & Encyclopaedia available at http://www.wordiq.com/definition/bodhidharma/html

105. Tang Yun, "India dreams of being a space giant", *Beijing Review,* 06 February 2003.

106. Dr. V. Siddharta, "Military dimensions in the future of the Indian presence in space", *USI Journal,* Vol. Cxxx, No. 540, Apr–Jun 2000, p. 253.

107. Air Marshal I.W. Sabhaney, "Possible countermeasure against satellite reconnaissance" *USI Publications,* September 1980.

108. Steven Lambakis "Space Control in Desert Storm & beyond", *Orbis,* Summer 1995, p. 427.

109. "Military benefits", Strategic Digest of IDSA, Mar 2004, p. 451.

110. Ibid.

111. For the entire story on IAF efforts to procure PRB/ELT, see "Wake-up call: locating IAF planes will be easier from '05", *The Times of India,* 26 May 2004.

112. See "An Aerospace Command", *Vayu* (3/2003), p. 16. Five years hence, the status quo continues with no aerospace command being formed. In fact the Government of India has refused the formation of an aerospace command.

113. The previous government had also turned down the IAF's proposal citing an "inter-ministerial turf war", see India's national newspaper, *The Hindu,* 19 September 2005.

114. Statement by then Air Chief Marshal S. Krishnaswamy, *PTI Report,* 06 October 2003 at URL www.newindpress.com/Print.asp?ID+"+val or Rediff news at www.rediff.com/news/2003/oct/06iaf.htm

115. Therasa Hitches, Vice President, Center for Defense Information, Developments in Military Space: Movement Towards Space Weapons, October 2003, p. 9. The author has premised her opinion on a source which has little or nil credibility and is not even known to be a registered journal in India.

116. *Yorkshire CND* briefing June 2004, "Keep Space for Peace", Part-2 at cndyorks.gn.apc.org/yspace/overview.htm

117. A comparative analysis of the *Yorkshire CND brief* and *Daily Times Pakistan,* report by Iftikar Gilani, "India Building Nuclear Attack Platforms in Space" clearly indicates that Yorkshire CND's brief is influenced in no small measure by misinterpretation and downright misrepresentation of facts by Daily Times Pakistan. See site edition at www.dailytimes.com.pk/default.asp?page=story_7-10-2003-pg_1_7

118. See Jeffrey Lewis, "What if Space were Weaponised? Possible consequences for crisis scenarios", Chapter 5, p. 30.

119. Hu Xiaodi, "Prevention of weaponisation of outer space and the work of the CD", disarmament conference, Geneva, 07 Feb. 2002.

120. Tang Yun, "India dreams of being a space giant", *Beijing Review,* 06 February 2003.

121. Unintended consequences: proliferation in South Korea", *Strategic Forecasting,* LLC, 2001, from www.orbireport.com/Columns/01-03-06.html.

122. Report on Korea's national space programme (2000–2015), *Publication of National Science and Technology Council,* Republic of Korea.

123. Keynote speech at UNISPACE by Korean representative, available at http://www.un.org/events/unispace3/speeches/19kor.htm

124. Hong Yul Paik, "Space development in Korea", Towards Air & Space..., Rand Publications, *CF-177-FIAS,* 2003, Chapter 5.

125. "Korea to launch first military satellite in 2006", *Korea Herald,* 18 November, 2002.

126. Ibid.

127. Hong Yul Paik, "Space development in Korea", *Towards Air & Space...,* Rand publications, *CF-177-FIAS,* 2003, Chapter 5.

128. "Palapa-B", Boeing's site at www.boeing.com/defense-space/space/bss/factsheets/376/palapa_a.html.

129. Ibid.

130. "Indonesia to launch microsatellite in 2005", *GIS development.net* from www.gisdevelopment.net/news/viewn/asp?id=GIS:N_neorhaixby

131. IBR Supancana and Susetyo Mulyodrono, "Indonesian space policies and institution", National Institute of Aeronautics and Space, Indonesia.

132. Ibid.

133. "Malaysia to launch eight small satellites in orbit", *Utusan Malaysia,* 11 October 2001.

134. Statement by head of the delegation of Iran, Dr Mostafa Safavi Hemarni at UN from www.un.org/events/unispace3/speeches/2/irn.htm

135. BBC news, "Teheran aims for satellite launch", 5 January 2004 from http://news.bbc.co.uk/2/hi/middle_east/3370143.stm

136. "Iran says will launch first satellite in April", *Space Wire,* 2 September 2004.

137. FAS, "Iris and Iran's emerging space programme" from www.fas.org/nuke/guide/iran/missile/iris.htm

138. "First satellite launch next year", *Iran Daily,* 5 October 2004.

139. Clifford Beal, "Iran's Shahab-4 is Soviet SS-4, says US Intelligence", *Jane's Defence Weekly,* Special Volume/issue: Vol. 031, 17 February 1999.

140. FAS, "North Korea space guide", www.fas.org/spp/guide/dprk/html.

Chapter 7

Examining Space-Based Options
for National Power

Space capabilities are becoming absolutely essential for national development, economic well-being, commerce, and everyday life, besides becoming a crucial component of successful military operations. It is generally well known and accepted that space has emerged as an essential component in furthering a nation's comprehensive national power. India has a robust civil space programme which is essentially geared towards scientific and development goals. As we move towards greater development, utilisation of space for economic and developmental purposes is likely to increase, and as dependence on space assets and systems increases, the concurrent vulnerability of our country to hostile action seeking to destroy, degrade or deny our space capabilities so painstakingly built over the decades would increase. India's dependence on space for vital economic purposes has been growing rapidly, hence any serious damage or degradation would have a major negative impact on our nation's well-being.

The lessons of history, on the other hand, are clear that wherever serious threats to national economic interests arise, military force would be necessary to protect them in the best manner possible. Military organisations have evolved as instruments of national power to protect national interests and investments. This generates the rationale for military involvement in space; besides the fact that space-enabled capabilities are the core of Revolution in Military Affairs (RMA) aimed at enhancing terrestrial military capabilities and national defence. The kind of modern precision warfare witnessed during the Gulf War is largely a by-product of this RMA which is aimed at combining the cumulative potential of air and space forces in terms of Intelligence Surveillance Reconnaissance (ISR), communications, navigation, etc. for providing information dominance vital to nuanced application of force which in-turn enables decisive war-winning effects.

Military Uses of Space

The military uses of space expand with every passing conflict as emerging technologies afford greater exploitability of the environment for pursuance of military activities. Until the last conflict, however, the uses were largely of a pacifist military though "non-weapon" nature. Space-based assets were mainly aimed at 'force-enhancement' missions like observation, communications, navigation, meteorology, etc. which allowed terrestrial military forces to conduct military affairs more efficiently. Thus most military space missions were auxiliary to other more direct military activities. The capacity to deliberately cause damage to another party is not the main criterion for attributing a military character to satellites. Most present-day satellites (excluding ASATs) affording military capabilities or performing military functions are incapable of directly destroying or damaging another country's property. Apart from 'Early-Warning' satellites which have a clear-cut military role, most of the other military activities can also be performed by civilian satellites and vice-versa. For example, civilian Earth-observation satellites are used for military remote sensing, civilian (even commercial) communication satellites have been known to carry military transponders, military navigation satellites have overwhelming civilian users, etc.

However, as military as well as commercial reliance on satellites grows, so too has the awareness that space-based assets are centres of gravity which are likely to be targeted in war. This in turn has fuelled the quest for development of techniques for protecting one's assets in space as well as denying an adversary the use of space. Thus, while up to the last conflict involving space, space systems were mainly focused on force-enhancement missions the present focus has shifted to controlling the realm of space for one's own benefit while denying it to the adversary.[1] The accent on military utilisation of space is gradually shifting beyond enhancement of military force capabilities to control of the environment and actual application of military force "in, from and through space".[2] The above trend is evidenced in the quest of space-superpowers like the US embarking on programmes aimed at space control and space projection.[3] Some of these include programmes like the Experimental

Satellite Series (XSS) which seeks to use small satellites to manoeuvre around other satellites in order to inspect, service or attack. They also include Kinetic Energy Anti-Satellite (KEASAT) systems, Directed Energy programmes as well as 'Counter-Space' initiatives like the Counter Communications System (CCS) aimed at disrupting satellite-based communication used by an enemy for military purposes. The first of such CCS system was delivered to the US's 76th Space Control Squadron in the year 2004.[4] Apart from the above, a "space based interceptor test-bed" programme is also under way to develop and test space-based miniature missile defence interceptors. The Pentagon's Missile Defence Agency has already provisioned budgetary allocations for the same.[5] The concept broadly envisages a limited constellation of space-based interceptors of 50 to 100 satellites offering a thin boost/ascent defence against ICBMs and a multishot mid-course defence against medium to intercontinental range missiles. The agency's plans call for the first contract to be let out in 2008, the first intercept tests by 2012 and "a constellation production decision" by 2014.[6] From the foregoing it is amply evident that space-based systems are presently in the process of transition from an era of militarisation to weaponisation.

Military Uses for Nascent Space Powers

It needs to be borne in mind that the aforementioned transition is applicable only to nations like the US. Its next closest rivals, the Russians and the Chinese are yet to embark on any serious weaponisation programmes. This is mainly on account of the prohibitive costs and technological challenges involved rather than lofty ethical considerations. The Russians inherited the entire range of capabilities for force-enhancement missions from the FSU. However, since the 1990s, its capabilities have been severely degraded due to funding problems. As of 2004, Russia maintained military space programmes only in five areas of early warning, optical reconnaissance, communication, navigation, and signal intelligence.[7] With regard to ASATs, the FSU was the only country that developed and operationally deployed an anti-satellites system (ASAT), designed to attack satellites on low-earth orbits. However, the present RF is not known to have any operational ASAT systems.[8] As for the Chinese, though they are the undisputed leaders in Asia in relative terms vis-à-vis the US, their capabilities are nascent. And for ASATs, speculation on the subject is rife and China is known to be actively pursuing such capabilities though it has not presently succeeded in its efforts.[9] Other countries with known space-based force enhancement assets in operation include France (Helios image intelligence satellite and the Telecomm-2 communications satellite), Italy (Sicral communications satellite), Spain (Hispasat communications satellite), Britain (Skynet-4 communications satellites), Israel (Eros and Ofeq imagery intelligence satellites), India (TES photo-reconnaissance satellites); Japan (commercial Superbird communications satellite system and Information Gathering Satellite); and South Korea (Kompsat-1 remote sensing satellite).

Thus apart from the US, most nations are yet to progress beyond rudimentary military space capabilities and force enhancement missions.

Space-based military Force Enhancement

In view of the above a brief exposition of the capabilities afforded by space systems for force-enhancement would be in order and the same is briefly undertaken.

Early Warning Satellites

These are used to monitor enemy territory for military activity such as missile or satellite launches, missile tests as well as nuclear detonations. Space-based sensors are capable of detecting ballistic missiles almost immediately after launch and provide maximum warning time for retaliatory or counter action, thus they constitute a vital component of EW systems.

Observation/Intelligence Surveillance Reconnaissance (ISR)

Their primary contribution is to enable situational awareness by providing information about a multitude of military activities by generating high-resolution images of areas of interest, monitoring changes, strengths and locations of forces, etc. This includes the important sub-sets of Imagery Intelligence, Signals Intelligence, Electronic Intelligence, etc. Satellites fitted with requisite sensors are also used for ocean surveillance.

Communications

These enable military commanders to exercise command and control over their forces and to receive real-time information about the progress of a campaign or about possible enemy actions to a degree that was previously unknown. Apart from these space and terrestrial sensors involved in ISR, navigation, etc. generate enormous amounts of data. The transmission of this and other data for military purposes needs reliable and secure communication which is provided by communication satellites.

Navigation Satellites

These are used to provide accurate targeting, positioning and navigational location information to users for strategic, operational and tactical requirements. These help military forces to precisely manoeuvre, synchronise actions, locate and attack targets as well as locate and recover stranded personnel and many other actions. They have profoundly improved the military efficacy of reconnaissance, accuracy and safety of weapon delivery platforms as well as weapon delivery itself, in addition to accurate deployment/re-deployment of military forces for delivering nuanced weight of effect, etc.

Meteorological Satellite

These satellites provide information about atmospheric water vapour, temperature and other weather phenomena. They are instrumental in determining the most appropriate moment for attack; they also monitor meteorological conditions during flight over target areas as well as providing real-time weather information over the area of interest. They provide data about cloud cover so that satellite reconnaissance missions can be planned efficiently.

Geodetic Satellites

These produce maps of the earth by using photographic and radar techniques. They also provide data about the earth's gravitational and magnetic fields which enable trajectories of ballistic missiles to be predicted accurately and are essential for the guidance systems of cruise missiles.

Advantages and Spin-offs

Apart from the aforementioned apparent capabilities, numerous other military advantages and spin-offs are obtained, some of which are:

Battlefield Transparency

The above would be enabled by a coherent mix of space, air and surface-based sensors complemented with multi-sensor data fusion capability to enhance 'situational awareness' at all levels. This capability is vital to combat decision making and force as well as weapon employment.

Freedom of Operations

This situational awareness in conjunction with the enormous information afforded by air and space platforms would provide vital inputs for successful prosecution of military operations. This in turn would ensure that friendly forces are used discriminately for delivering weight of effect at the right time and right place thereby enabling effective operational employment. This information would also expose friendly forces to lesser risks by navigating them safely on to the target. This freedom of action would create operational opportunities for own forces while conversely limiting the adversaries.

Persistent Over-watch

Space offers the potential for safe persistent over-watch over designated areas of interest. Apart from the vertical depth and strategic breath of vision they would offer, enormous intelligence inputs would be forthcoming which could ensure that the adversary is under constant surveillance thereby reducing his opportunities to 'surprise' and seize the initiative.

Deterrence

The above in turn would also endow effective deterrence. Persistent over-watch would continuously monitor illegal cross-border movements and transaction thereby reducing border porosity. In fact, even during sensor switch-off periods the psychological impact of continuous multi sensor detectability[10] shall deter potential mischief and misadventure. In peacetime, it can promote regional stability by its persistent over-watch capabilities and during crises and wars it would enable rapid response across the spectrum of conflict by its ability to deter, contain, resolve or engage and win.

Decision Making

Space-based inputs augmented by conventional methods could be used for creating information databases on adversaries during peacetime, so that effort and resources are not disproportionately expended in acquiring basic inputs during 'immense of hostilities'. In addition, during crises and wars they would enable effective decision-making by providing accurate and real (or at least near real-time) inputs vital to military decisions at strategic, operational and tactical levels.

Other Spin-Offs

In addition to the above, a number of other advantage accrues to the military from space which is limited only by the ingenuity of the user and his 'employability-awareness' of the subject. For example, prevailing MetSats can be used for enhancing battle-field employability and manoeuvre, existing local user terminals for search and rescue can be applied for rescue of stranded combatants, ejecting aircrew, etc.

Imperatives of our Defence Needs

From the foregoing it is apparent that space-based systems provide vital capabilities to successfully execute national military strategy in addition to the overall grand strategy and have the potential to be used across the range of military operations at the strategic, operational and tactical levels of war in order to accomplish national security objectives. Secondly, information derived from air and space platforms would be vital for success in conflicts. Hence it would be imperative to attain a certain modicum of 'information-dominance' in order to complement our conventional capabilities. Thus we need to enhance our conventional military prowess by harnessing available space capabilities and potential so as to comprehensively reciprocate to the spectrum of warfare being directed towards us and also limit (if not deny) our adversaries the opportunity to offset conventional military superiority by resorting to threats of WMD, or other forms of unconventional warfare. There exists an emergent need for examining the options afforded by space in order to address the following aspects.

- Securing of our space and terrestrial assets and thereby ensuring uninterrupted national development.

- Coordination of military requirements and development of military space capabilities.
- Integration of space and conventional military capabilities.

Requirement of an Agency for Coordinating Space-based Military Affairs

Space offers a number of war-winning capabilities like near instantaneous communications, continuous surveillance and highly accurate positioning. These capabilities provide a decisive advantage to the military. India has formidable civilian space capabilities and dismal military space capabilities. Over the years India has built up adequate capability in space technology and our space assets are being exploited efficiently by the civil sector for a number of applications but the military use of space by India has been minimal. Present Indian efforts are grossly inadequate and uncoordinated in the absence of a central coordinating body. Indian military use of space is limited to procuring imagery from a single military satellite, the Technological Experiment Satellite. India is yet to integrate its formidable civilian capabilities into its war-fighting machinery or at least take steps to protect its assets in space. There has been no single coherent agency for managing our military space programme resulting in sub optimal and inadequate project management and exploitation of our extant space capabilities. This lack of coherent leadership in military space application has resulted in the significant gains being under-utilised in a military context. India ranks fourth in the world with regard to both national airpower and space capabilities; however, the absence of a central coordinating body like an Aerospace Command has led to a situation wherein conjoint exploitation of the mediums of air and space is the lowest in the world as evidenced in Figure 7.1.

Thus the requirement of a centralised agency aimed at coordinating national military space effort in support of national military space objectives is imperative and inescapable. While no attempt is made herein to comprehensively enlist all the tasks such a proposed agency would fulfill, broadly the following requirements would be met.

- It would serve as a single point of contact for all space applications and support, responsive to all security agencies.
- It would serve as a nodal agency for space system/capability

Composite Aerospace Exploitation Across the Globe[4]

Country	Airpower (No. of Aircraft–year 2003	No. of Military Satellites–year 2003	Degree of Integration
USA	16,038	105	Highest
Russia	9,211	144	High
China	9,812	06 Dedicated, dual–use not known	Medium
India	1,813	01	Nil
Japan	1,754	02 launched (in orbit 02 failed	Medium
Israel	1,038	02 in orbit	High
		01 dual use	
		02 failed	

Figure 7.1

demand, acquisition, procurement, etc. from external agencies (both civil and commercial).

- It could be entrusted with the responsibility of coordinating the requirements of various users, consolidating all existing and proposed space support applications, recommending compatibilities and feasibilities of such applications and scrutinising support requirements.

- It could be tasked to examine and recommend suitable defence-oriented systems/sub systems, deployment of such systems and subject to approval pursue implementation of such projects.

The Case for an Aerospace Command

In view of the foregoing it would be essential to explore options available for choosing a suitable agency before arriving at conclusion and hence the same are explored. The main options rampant in debate and discussion have been examined below. Following an objective appraisal, the choice of a suitable agency would automatically be forthcoming.

An Integrated Strategic/Space Command

The above has been touted as an option on the basis of the following:[12]

- That "the elements linked to space assets would be focussed primarily towards strategic objectives, both in defensive and strike roles. These ingredients are essential for the basic functioning of an integrated Strategic Nuclear Command (SNC) and since they would form the basic input source for SNCHQ's and National Command Centres they should be placed under the operational command of the CDS".

- Drawing on lessons of the US and the RF's restructuring in that the US has undergone a restructuring and integrated its Strategic Air Command and Space Command. Similarly, the RF has also undergone a restructuring by dismantling its SRF and merging the ADF with the Air Forces and placing them under unified general staff. The above analogy implies that India should also be doing the same.

ISRO/Civilian Space Command

- In view of ISRO's all-pervasive proven record in space, apart from the above choice it would be the prime choice, hence it should form the space command.

Examining the Strategic/Space Command Option

The idea that the elements linked to space assets are primarily focussed towards strategic objectives is basically flawed. It is common knowledge that military space capabilities contribute to all levels of military activity and conflict. Space assets like airpower assets are inherently complex instruments of warfare and impact all the three levels of strategy, operations, tactics not only in terms of objectives but also in terms of mission, effects produced, platforms, munitions, etc. hence attempting to force-fit them into water-tight strategic compartments would amount to underutilisation of assets. By this logic, all the Air Force elements should also be under the above-mentioned umbrella. A balanced holistic approach would dictate that current national military space objectives require supporting a strong, stable, and balanced national space programme that serves our goals in national security, foreign policy, economic growth environmental stewardship, and scientific and technical excellence. The point is that unique circumstances demand unique organisational doctrines though at the environmental level, doctrines would largely be similar. Hence Indian military space endeavours would have to be primarily focused on all-encompassing strategic, operational and tactical missions rather than just a particular set of missions which would amount to gross underutilisation of potent capabilities. Secondly, a brief perspective into the evolution of space as an enabler of military operations would be instructive and hence is undertaken. Military space capabilities have generally evolved through four stages historically. In the first stage or era, it was focused on meeting the strategic objectives of the Cold War (similar objectives are absent and inapplicable in our unique context). By the end of the Cold War, the focus shifted to tactical and operational utility of space systems in enhancing conventional military capabilities (which is more applicable in our context). Towards the end of the 1980s, the focus was on integrating space into military operations (again applicable in our context) and following the successful experiment of the Gulf War came the era of expansion beyond support into actual control of the environment for military and economic might preservation by harnessing the strengths of the environment for producing unique war-winning effects. This dictates the rationale for the organisational merger in the US.[13]

The US reorganisation experience reveals that while its Air Force space command was formed in 1982, its Naval command the next year, a Space Command in 1985 and an Army space command in 1988, the rest of them either stagnated, or maintained status quo, or were disbanded altogether whereas the Air Force Space Command gradually grew in responsibility and resources. The US space command was disbanded in 2002 while the Navy and Army space command ceased to exist.[14] Though the other service commands and a unified command responsible for war fighting with space forces have existed for over 18 years, responsibility and stewardship of space forces has been an Air Force preserve since 23 years. The Air Force thus has served and continues to serve as the longest and best steward in furthering their national space capabilities though it never sought stewardship for itself, nor was it assigned the role of space-steward, stewardship automatically devolved on to the Air Force. As of now, the US is set to combine all of the forces that the Air Force provides to USSTRATCOM—intercontinental ballistic missiles; space forces; information operations; intelligence, surveillance, and reconnaissance (ISR); and global strike—into a single component, which will be called Air Forces Strategic Command (AFSTRT).[15]

While the idea behind organisational restructuring in the US is clear, the reorganisation of Space forces of the RF needs to be viewed in a uniquely (modern) Russian rather than arcane Soviet perspective. Following the collapse of the Soviet Union, the loss of integrated Soviet military structures, financial crises, resource crunches and a host of other factors, comprehensive reforms in Russian air power were effected to make them more efficient and cost-effective. In tune with prevailing doctrines, and the US experiences in the Gulf War, etc. the accent on efficient use of airpower shifted on to centralisation of all aerospace assets within a central agency like the Russian Air Force; the *Voyenno-Vozdushnyye Sily* (VVS). Consequently, independent services of the RF like the Air Defence Forces, the *Voyska Protivovozdushnoy Oborny* (VPVO) as well as branches of the independent services like the RF Army's Aviation Branch, the *Aviatsiya Sukhoputnykh Voysk* (ASV) were disbanded, amalgamated and subordinated into the VVS in 1998 and 2003 respectively.[16] The current reorganisation would indicate (and advocate) that the responsibility be entrusted to the IAF since, firstly Air Defence (and by typically Russian logical extension 'aerospace defence') in India is the axiomatic responsibility of the IAF, secondly its space systems were largely an Air Force preserve, with the Navy and Russian Ground Forces having no apparent major space responsibility.[17] The present merger reinforces the premise of centralised control of aerospace assets within a parent organisation like the air force.

Finally, focussing largely on strategic applications would amount to grossly underutilising the potent capabilities of space. In such an arrangement, space's contribution to tactical operations of air, land and sea forces would be significantly reduced. This would severely affect the tactical operations of the three services which certainly is no commendable manner of optimally utilising space.

The CDS/IDS/CIDS Option

The CDS is yet to evolve into a full-fledged instrument of national power, hence bestowing an important task like integrating space on conventional military elements without the wherewithal would certainly not be a good option for the present or even in the near foreseeable future. Secondly, the question of integrating space demands immediately and the CDS apparently is in no position to fulfill the same. The CDS in its present avatar has no operational role and is merely a coordinating agency, therefore it not expected to be able to wield the requisite influence or effect required in operationalising space. Lastly, space needs to be utilised as an integrated aerospace construct with air and space capabilities and not in isolation. The CDS is surely in no position to fulfil this.

The ISRO Option

This option is not viable in view of the following:

- ISRO has a dedicated civilian agenda and is largely successful in productive commercial endeavours. Re-orienting ISRO's space efforts to a military or semi-military agenda would certainly entail sacrifices in terms of diversion of resources, manpower, efforts, etc. which certainly would lead to a depletion of its overall capabilities.

- Nevertheless, ISRO would have the option to invest more funds and demand more budgetary allocations for dedicating resources to military purposes, training of civilian personnel in military nuances, training civilian personnel to perform military jobs and the whole gamut of requirements to shift from a civilian to a military agenda which certainly would be enormous. All the above and many more unlisted tasks could already be undertaken by a distinct military institution. Thus the wisdom of such an option appears eminently questionable, if not absurd.

- Lastly, the question of legal appropriateness of civilians performing military tasks would need to be addressed. The complications are enormous. As per the Laws of Armed Conflict, civilians are not authorised to take a "direct part in the hostilities". Persons who commit combatant acts without authorisation are unlawful combatants and are subject to criminal prosecution. The extent to which the above option would appeal to the hugely law-abiding scientists of ISRO would certainly endorse the absurdity of the proposal.

A Distinct "Space Force" Option

Independence is not appropriate for space today. The Air Force was established as an independent force when air power had at least reached adolescence—only after combat-tested technology, doctrine, and leadership were well established. Military space is still in its infancy, with no unique mission, untested doctrine and personnel, and unfinished technology. Military space capabilities contribute to all levels of military activity and conflict but have yet to evolve into a full-spectrum, war fighting force.[18] The US experience suggests that space should be allowed to mature within an established parent organisation like the Air Force to determine whether it can develop and refine a unique war-fighting capability.

Secondly, if by summer of 2005 the lone aerospace super-power still believes that US military mission in space has not sufficiently evolved to warrant the establishment of a separate military service for space operations, would it not be premature or rather outrageous to believe that with a single military satellite and a few military space capabilities envisaged a separate space force can be sought for?

Thirdly, the argument for an independent space force does not altogether dismiss the aerospace concept. As and when conceived it would be sustained by the logic of enduring aerospace capabilities and emerging space strengths. Most nations have ancillary air support organisations supporting their Armies and Navies. In fact, the Indian Army's AOP division is larger than over a hundred air forces of the world. Thus as and when military space capabilities evolve beyond terrestrial support under the aegis of air forces to independent forces to independent forces, certain capabilities would experience full time migration to space and some would continue with the air forces.

Establishment of such a force would demand enormous space and military investment. The entire administration, logistics, operational and technical capabilities and human resources would have to be raised afresh. Such a force would lack the in-depth war fighting perspective and experience inherent in the other services. All-in-all, a total duplication of efforts or reinvention of the wheel, which would not only be a wasteful and costly endeavour but may also turn out to be counter-productive. Lastly, the development of an independent space force might signal to the rest of the world that we intend to weaponise space and could result in unwanted international pressures and even sanctions.

The Available Option

In view of the logic of aerospace continuity, the global trends, the evidence of recent conflicts, etc. it is obvious that space ownership is decided by factors like doctrinal similarly which enable smooth harnessing of capabilities in transforming conventional military prowess, operational proven concepts which allow assignment of time-tested roles and missions to space, and similarity of environmental characteristics apart from contiguity which allow integration and operationalisation of capabilities.

While there is no disputing the fact that space support would enhance capabilities of every service, what distinguishes the Air Force beyond enhancement of its capabilities is the fact that subsequently some of its roles and missions would partly migrate into space as evidenced by the migration of high-level reconnaissance from aerial to space-based platforms right at the beginning of the space age to the present wherein many missions like meteorology, navigation, etc. have already migrated into space. Such a migration would cause enormous complexities in terms of 'aerospace management' and the only agency which could co-ordinate this vital aspect would be the Air Force.

It should also be borne in mind that a fundamental principle of organisation is that the resource management chain is not built around specific assets or around specific missions of geographic areas. Rather it is organised around the homogeneity of its force structure. Air and space capabilities are complementary rather than competitive, hence the sensible option would be to integrate the above homogenous capabilities rather than pit one against the other by isolating them prematurely into separate forces or attempting to force-fit them into disparate structures. Such homogeneity for enabling optimum management and utilisation of space resources would in the present be an Air Force preserve.

Last but not least, in the eventuality of Indian space assets being targeted by hostile elements like parasitic ASATs, airborne lasers from Boeing-737s, or air-launched projectiles, the only response option available on the spot for security and protection of assets would be the IAF. The only element in our military arsenal which could retaliate or encounter the aforementioned threats would be fighter aircrafts. Conceptually, the ASAT launch pads could be attacked by fighter ac with stand-off munitions supported by air or space based positioning. B-737s bearing ASAT lasers could be shot down by fighter aircrafts. Air-launched ASAT options are also not impossible considering the fact that until a congressional ban on ASATs, the Americans had experimented with miniature homing vehicles mounted on F-15 fighter jets, the Russians are known to have altered Mig-31s for an ASAT carrier roles and present-day research indicates that fast fighter jets like the Eurofighter Typhoon could easily be used to launch microsatellites into orbit.[19]

Finally, in the eventuality of satellites providing remote sensing, navigation, meteorological input, etc. being destroyed, the loss could be immediately compensated by substituting an aerial platform to perform the same role. Thus the continuity and tempo of operations would be maintained by an integral mix of air and space compensating for each others deficiencies rather than working in isolation.

Implications

It is against this context that we in India need to address the role of space in our national defence and security. The Ministry of Defence had confirmed to the Standing Committee on Defence (1999–2000) of thirteenth Lok Sabha that "it has been proposed to impart a defence orientation to our successful space programme by including surveillance sensors, communications and navigation satellites". Additionally, the Standing committee on Defence reiterated their recommendations in their 14th report and desired that an Aerospace Command project be taken up seriously to ensure that India comes on the global space map.[20]

Two issues merit attention here; one across the globe stewardship of space rests with Air Forces. Additionally, apart from the continuity of the vertical expanses, the logic of seamless air and space mediums, the doctrinal analogies, etc. the above time-tested pattern would obviously be the ideal option for the country because between the choice of rational continuity and fantastic discontinuity, the prudent choice would obviously rest on continuity. Radical experimentation by departing from time-tested conventions may be an un-affordable option for a cost and techno-intensive arena like space. This no doubt is the reason that the Standing Committee on Defence has emphasised the importance of an Aerospace command in the Indian Air Force. Secondly, the spectrum of warfare directed against us is expanding and hence, it is only natural that our quest for befitting response options reciprocally expands, and if the tremendous impact of space enabled capabilities in the last few conflicts is taken as an example, the pathway appears clear, especially in view of the capabilities afforded by our civilian space programmes. The issue is not to follow the American trail-blaze on militarisation/weaponisation of space, but decide how best to optimally harness prevailing space capabilities within our own uniquely different security, economical and political context. The underlying endeavour would demand articulation and identification of the key issues, their placement in the larger context of a changing environment, and definition of new options for the exercise of space power in the pursuit of national interests within the scope of our unique capabilities and limitations.

Examining Contemporary Cases Worldwide

In addition, a brief look at the global trend (shown in Figure 7.2) in space stewardship would be in order. ⊃

Global Trend in Space Stewardship

Country	Space Steward	Remarks
USA	Air Force	90% of assets contributed by Air Force.
Soviet Union/Russia	Air Force/Space Force	Contemporary Soviet missions of SRF, PVO in an Indian context are presently executed by the IAF. The space force formed in year 2001 is an attempt to consolidate dwindling capabilities of SRF, PVO and Air Force Space Systems.
China	CMC+PLAAF	Chinese military modernisation and "informational-isation" concepts are premised upon pursuit of space-enabled RMA by a modern Air Force.
Japan	Air Self Defence Forces	
Israel	Air Force	Israeli Air Force in the process of converting into air and space arm.
S. Korea	Air Force	Korea Aerospace Research Institute established for coordinating all air and space development activities.
Indonesia	Air Force	In response to violation of their airspace by 05 USAF F-18 fighter ac in July 2003.

Figure 7.2

Thus it is conclusively evident that Air Force are the global choice for harnessing space capabilities in support of national security objectives. No nation in the world is known to have divorced space from their respective Air Forces. The common capabilities of space and air assets need to be optimally exploited with correct prioritisation and balance. Hence given the choice of time-tested and proven continuity as opposed to radical discontinuity the obvious choice would devolve upon Air Forces.

Conclusion

Apart from the wisdom of time-tested continuity as opposed to radical discontinuity in proposing the Air Force as the lead agency for conduct of space operations, it is common knowledge that overall responsibility for defence of the nation from all airborne threats and for all military offensive operations employing the vertical medium rests with the Indian Air Force (IAF). Technological advancements have today not only opened the vertical expanse beyond the atmosphere for exploitation but have also fused the mediums of air and space to the extent that apart from allowing the migration of certain aerial capabilities like observation, reconnaissance, navigation, etc. further up in space it has also integrated air and space capabilities into terrestrial war fighting to the extent that precision war fighting has become the most sought-after norm rather than the exception. Technological evolution of air-breathing platforms like aeroplanes into hyper planes designed to exploit both the aerial and space environments would demand supporting infrastructure for conjoint applicability of capabilities. Even from a defensive paradigm, air and space are complementary to each other, considering that both as complementary components of an air defence system would compensate for each others' deficiencies in countering threats of systems like ICBMs which transit and manoeuvre through both the mediums of air and space. Thus a requirement of integration of conjoint aerospace capabilities into terrestrial war fighting for concentration of force/effect would dictate that the command and control of the proposed Aerospace Command needs to rest with the IAF whereas the overall structure could be a joint effort of the three services and civilian scientific bodies.

Notes and References

1. Literature on the above shift in focus is in abundance, however, for a brief overview refer to compilation of quotes on the subject "The Final Frontier" by Michael Katz-Hyman, Stimson Center in *Arms Control Today*, November 2004, p. 13.

2. See "Report of the commission to assess United States National Security Space Management and Organisation", 11 January 2001, p. 16.

3. See Michael Krepon, "Weapons in the Heavens: A Radical and Reckless Option", *Arms Control Today*, November 2004, p. 10

4. See Jeffrey Lewis, "Programs to Watch", *Arms Control Today*, November 2004, p. 12. Also Nicole Gaudiano "USAF seeks weapons for counterspace capability" *Defense News*, 25 July 2005, p. 44.

5. Gopal Ratnam, "Killers from Space", *Armed Forces Journal,* June 2005, p. 24.

6. Ibid.

7. Pavel Podvig, "Russia and Military Uses of Space", Stanford University, Center for International Security and Cooperation available at http://russianforces.org/podvig/eng/publications/space/20040700aaas.shtml--

8. Ibid.

9. The US Department of Defence, "Annual Report on the Military Power of the People's Republic of China" July 2003, p. 36 claimed that China was developing killer microsatellites based largely on a January 2001 Hong Kong newspaper article. However, Gregory Kulacki and David Wright of the Union of Concerned Scientists traced the story to a website run by a self-described 'military enthusiast' named Hang Chaofei who ran a Chinese language Internet bulletin board filled with crude illustrations and "fanciful stories about secret Chinese weapons to be used against Americans in a future war over Taiwan". For details, see Greogory Kulacki and David Wright, "A Military Intelligence Failure? The Case of the Parasitic Satellite, Cambridge, MA: Union of Concerned Scientists, August 2004, p. 3.

10. It is common knowledge that space-based sensors are most effective when used in conjunction with other air-borne and ground-based sensors like aircrafts, UAVs, Surveillance Radars, etc. and not to their exclusion.

11. The number of aircraft has been sourced from "World Military Aircraft Inventory", *Aviation Week and Space Technology,* 13 Jan, 2003, pp. 257–76 and generally corresponds to equal figures in SIPRI's and Military Balance. However, military satellite estimates with respect to US and Russia differ in publications ranging from SIPRI's, the Teal Group, etc. to the US Air Force Association's Space Almanac, etc. Nevertheless, most publications are in common agreement regarding the estimates of Asian military spacecraft.

12. See Rear Adm A.P. Revi, "An Integrated Strategic/Space Command Option", *Indian Defence Review,* Jan–Mar 2005, Vol. 20(1), p. 87–90.

13. A series of studies was conducted in the late 1970s and early 1980s to begin studying ways to improve organisational structure for prosecuting space operations. They led to the conclusion that an operational space command was required for the Air Force to expand its potential in space. The Air Force Space Command gradually grew in responsibility and resources thereby validating the doctrinal premises. At the outset, its mission was confined to operating missile-warning satellites and sensors, and conducting space-surveillance activities. In 1985 it assumed satellite command-and-control responsibilities. In 1990 the space-launch function, as well as the responsibility for associated launch facilities and down-range tracking sites, was transferred to Air Force Space Command from Air Force Systems Command. And by the millennium ICBMs were transferred to the AF Space Command. The move of integrating AF Space Command Strategic Command were aimed at

better utilisation of extant capabilities and not reorganised because of AF failure in space stewardship or because the Americans belatedly realised their mistake as implied.

14. J.R. Wilson, "The ultimate high ground", *Armed Forces Journal*, January 2004, p. 29.

15. Gen. John P. Jumper, USAF Chief of Staff, to Adm James O. Ellis, Jr., letter, 23 February 2004, http://www.55srwa.org/0403/04-03011454.html.

16. See Marcel De Haas, "Russian Security and Airpower 1992-2002", Frank Cass, New York, 2004. Chapter 3, pp. 100–102 and 112–118.

17. In the Soviet era when its military space activities touched its zenith, the now disbanded Strategic Rocket Forces (SRF) held primary responsibility for launch of all civil and military satellites, the PVO (troops of air defence) were concerned with air and space defence, the Air Forces looked after space systems, search and rescue, cosmonaut training, vehicle recovery, etc. and the Navy and Ground Forces had no apparent major space responsibilities. For greater details refer to Nicholas L. Johnson, "Soviet Military Strategy in Space", *Jane's Publishing Company Limited*, London, 1987, Ch. 2.

18. Ralph Millsap, Dr. D.B. Posey, "Organisational Options for the Future Aerospace Force", *Aerospace Power Journal*, Summer 2000.

19. The conclusion of the paper published is simple: A 5-meter rocket weighing 2 metric tons could be fired from a Eurofighter in flight, powering a 50 to 70 kilogram satellite into orbit up to 700 km above the earth. For details, see Tom Kington, "Italy eyes air-launched minisatellites", *Defense News*, 25 July 2005, p. 30.

20. "An Aerospace Command", *Vayu Magazine*, March 2003, p. 16.

Chapter 8

Proposed Space Road Map for Fulfilling Security Requirements

National security establishments would need to seriously consider induction of space assets into conventional capabilities to bolster prevailing capabilities and conduct traditional tasks more efficiently. Existing space assets would need to be exploited at the earliest and future projects would need to take into account national security requirements along with economic, scientific, social and other requirements. The military uses of space have evolved from early, tentative attempts at enhancing conventional military force capabilities to actually enabling the decisive outcome of battle in present times. The reliance, rather over-reliance on space-based systems like in the US basically stems from the fact that such systems can accomplish or enable accomplishment of military missions more efficiently, more economically and in a more technologically superior manner than could be achieved by any other means. Acquisition of space-based military capabilities globally has become the most sought-after norm rather than the exception.

At the same time it also needs to be understood that space is not a substitute for all forms of military capabilities, or equally important, a panacea for all information voids or military inadequacies plaguing our national security concerns. Uses of space-based systems are constrained by exorbitant costs, high levels of technology, etc. Satellites themselves are constrained by technological factors like predictable over-pass timings and orbital patterns, width of coverage, attenuation due to inclement weather, etc. Its utility and limitations in the information loop, applicability at the strategic, operational tactical levels must be understood in the correct perspective for optimal exploitation. It also needs to be borne in mind that arriving at present levels of military space utilisation has taken even an advanced nation like the US the better of over four and a half decades. Hence we would need to be extremely cautious in comparing our capabilities with those of such advanced nations, in pursuing capabilities and goals typical to their abilities and needs, and also in superimposing their advanced doctrines and strategies on our relatively nascent needs and capabilities. Defence strategy on space should be dictated by rational security needs and not the outer limits of what appears to be technically possible as in the case of super powers like the US.

Developing Space Capabilities for Defence

Nonetheless, it is crucial that we incorporate existing space capabilities, resources and operations in our security policies and strategy. Our unique circumstances and practical considerations would demand a deeply cooperative and even integrated approach to capacity building and resource utilisation by various segments of the government. It would be counter-productive to duplicate assets or to re-invent the wheel; affordability and optimisation would demand a cooperative evolutionary approach. India already has a fairly advanced programme for economic, social development purposes, the programmes could be modified to include space capabilities needed for security purposes. Space capabilities for defence could be built upon as an extension of this programme thereby enabling affordability and optimal utilisation of resources.

Space capabilities for defence would primarily need to be built upon two considerations. One, a balanced assessment of the military requirements vis-à-vis capabilities afforded by space and second, the coordination and integration of space into our conventional security apparatus. Prioritisation of requirements, technologies and capabilities would need to be objectively charted and pursued for development and deployment of capabilities. This would be essential for evolving a robust, credible and affordable programme which would be a mutually beneficial, cohesive and affordable. It could fulfil national defence needs and at the same time not drain or compete with the civilian developmental programmes. The structure could be designed to ensure that the civil and military aspects of the national space programme complement and draw strength from each other rather than compete with each other.

Civil to Military Transfer of Capabilities

We have significant experience and expertise in space and space-related technologies. Many of them would have dual-use applications while some of them would have to be further developed to support national defence goals. Civil and military space activities are complementary and no extra-ordinary 'budget-draining' effort is presently foreseen for technology transfer from civil to military space endeavours. For example, launchers make no difference between civil and military payloads. Similarly communication, navigation, imagery, meteorology and geodesic satellites have both military and civil applications. In fact, space-based imagery, communications, etc. are also available off the shelf as commercial products. It is expected and presently accepted, that economic (and perhaps political) considerations may limit some civil to military spin-offs. However, the compulsions of national interests would endorse the approach that capabilities for defence should not be divorced from the economic and commercial uses of space. These are regarded as challenges to be overcome jointly in the larger national security interests, rather than permanent obstacles in building up our military space capabilities.

While certain dual-use capabilities could be harnessed right away, other capabilities would demand dedicated efforts. Certain defence-specific technologies would need to be developed and progressed, and defence R&D could concentrate on defence-specific technologies that cannot be more easily developed within the ISRO framework. Joint effort by a central defence coordinating agency like an aerospace command, space agencies and defence technological agencies like DRDO could undertake development, planning and execution of such programmes.

Thus, keeping in mind the 'availability and affordability' criteria, presently available space technologies need to be dovetailed to meet present national security and defence requirements, and future requirements foreseen should be projected with due attention to costs, legalities and treaties in vogue, technical feasibilities, etc. In our unique context, military doctrine would need to be fit unto available technologies for the present, whereas, for future applications doctrine, requirement, technology and capability could synergistically evolve keeping pace with each other for optimal military exploitation and utility of space-based systems.

The present choice of military space applications could be a subtle mix of applications which need not be exclusively to military needs, but would be a natural adjunct of civilian applications, with a greater degree of autonomy, redundancy and security. Thus while present military requirements may deftly be included into the ISRO decade plan (1997–2007), future military requirements after due consideration could be projected well in advance to allow their inclusion and integration into the subsequent decade plans. This would enable synergistic evolution of national military space capabilities without disturbing the pace of our civilian space capabilities.

Road Map for Military Utilisation

The assets and capabilities required for a credible space programme aimed at national security in addition to civilian development would need to be carefully built up taking both the operational needs as well as the other issues of affordability, political correctness, cost, etc. into account. A significantly amount of capability and assets would need to be built up, with many of them created on a priority basis and/or strengthened even after they have been initiated. The needs and developments would need to be kept under constant review and monitored.

Road Map Categories

The list of technological advances and applications afforded by our extant civilian capabilities is long and impressive and there is an equally long list of the potential paths and options for exploiting these capabilities but fundamentally, the road map to the followed could be grouped into the following three categories.

Near Term (2007–2017)

- Application of available space capabilities for military force enhancement, i.e. using space for enabling more efficient use of conventional military capability.
- Institutionalising protection measures for securing space-based assets by measures like hardening, manoeuvrability, etc.
- Establishing organisation and infrastructural facilities to fulfil the above requirements.

Mid Term (2017–2027)

- Integrate space and conventional military capabilities to possess a comprehensive instrument of national power capable of delivering the collective might of military and space power on to objectives for furtherance of national objectives.

- Institutionalise measures for defending national territory and assets against military force application from space.

Long Term (2027–2037)

- Synergise operations and technology to go beyond incremental capability development and pursue better strategies for accomplishing national objectives.

Near Term (2007–2017) Requirements

In the near term, the focus would primarily be on building our capabilities by focussing on force enhancement missions and protecting our assets in space. The first aspect could be addressed by inducting space systems in terms of communication, navigation, observation capabilities, etc. into conventional military capabilities which would be within the scope of legalities, the realm of technologies available to us and also not be a drain on the exchequer. The emphasis would be on utilising the enormous information endowed by space systems to manage and conduct military affairs more efficiently. This would demand significant participation and cross-exchange of ideas and information between the military and civilian space agencies. The cross-flow of information would primarily be in terms of technological inputs and feasibilities from the civilian space agencies corresponding to operational requirements put forth by a central military coordinating agency after due in-house deliberation and consideration. The second aspect dealing with protection and security of our costly space-based assets built so painstakingly over the decades would also demand enormous cross-exchange of inputs and participation by both civilian and security agencies.

Building Capabilities

Reliable Communication Capabilities

In view of numerous factors ranging from immediate responsive communicability demanded by the extending reach of the IAF to the spread of the defence forces across the length and breadth of the subcontinent, the variety of terrain to the inherent limitations of ground-based communications systems, etc. there exists an emergent need for the armed forces to be enabled connectivity by space-based systems. The inherent virtues and efficiency of space-based communications would overcome prevailing problems of line-of-sight, attenuation losses, low reliability, and so on. Secondly, apart from enhancing C4I, it would also serve to efficiently transfer the enormous amount of data and information provided by space-based observation, navigation and other systems to the users in real-time or near real-time.

A military communications system in our Indian context would basically indicate military-unique systems owned, leased or dedicated to military purposes. These would encompass operations, control and employment of systems by the military. The systems could include the requisite number of transponders, frequency bands or even leased capabilities on civil or commercial satellites (with requisite legalities in place) rather than dedicated military communication satellites. Dedicated efforts would be prohibitively expensive and may amount to underutilisation in an era of multi-tasking. The essential factor distinguishing civil and military systems is that civil applications are aimed at reliable connectivity at most economical rates whereas defence applications would demand reliable connectivity even in the face of deliberate degradation, interference, etc. which are characteristic to combat operations. Requisite modification could be undertaken to balance and meet both the requirements. Broadly, the following capabilities would be demanded by the military from space-based systems:

- Assured, real or near-real time access, available on demand.
- Adequate redundancy to cater for emergencies.
- Adequate spectrum diversity and capacity to meet both current and emerging information requirements.
- Reliable, secure and redundant data relay system for information,

data transfer and transmission between satellites, airborne and terrestrial sensors, weapon platforms and other military systems.

■ Ability to efficiently undertake high data transmission of imagery, voice, test, etc.

■ Flexibility to match the dynamic operational environment.

■ Security and protection of systems, links, nodes, etc. against soft-kill by IW, etc.

Space-Based Observation (ISR) Capabilities

The term space-based observation has been used to broadly include missions of IMINT, ELINT, SIGINT, etc. which primary deal with space-based observation aimed at providing significant ISR capabilities. ISR requirements would entail integration and exploitation of prevailing and forthcoming advances in space-based sensors inclusive of capabilities on civilian application satellites like the IRS series, dedicated military efforts like the TES, and even commercial satellites like IKONOS. Such integration would not be limited to space-based ISR systems but also to corresponding air and surface-based systems because pursuing independent ISR capabilities would amount to defeating the purpose of such potent capabilities. Overall integration of capabilities would be essential to obtain a synergistic strategic, operational and tactical observation capability. For example, an integrated observation system with land-based ISR sensors would fulfil observation requirements of the contact battlefield up to 10 km as well as the intermediate battle up to 40 km; airborne ISR platforms could cover the area of interest of a Corp (which would generally be 300x300 kilometers). Areas even beyond which would be of interest to strike and other delivery mission aircraft would be covered by airborne as well as space-based systems. The broad requirements to be fulfilled for ISR capabilities would be as below:

IMINT

■ Systems providing 1 metre and even lesser resolution are presently available and would suffice for the near term, though repetitity and revisit factors need to be resolved.

■ Area of coverage contiguously needs to be extended to include areas where known threats exist.

- Availability of a judicious mix of sensor payloads like Hyper Spectral Imaging (HSI) and Synthetic Aperture Radar (SAR).
- Availability of better data processing and transmission capabilities.

ELINT/SIGINT

- While dedicated efforts for such capabilities would be most suitable, in the interim requisite payloads could be placed as 'piggyback options' on satellites to cover areas of permanent, near-permanent interest.
- Opportunities afforded by micro and nano satellites in fulfilling the requirements also need to be explored.
- Integrate space-based effort with conventional capabilities.

Ocean Reconnaissance

- This would be the only investment in near-term, which would be dedicated to purely military affairs; however, the same is considered inescapable on account of our expanding interests and reach of the Navy.
- In view of the enormity of the investment involved, the project may be undertaken phase-wise so as to spill over into the mid-term.

Space-based Positioning and Navigational Capabilities

The Northwest and Northeastern parts of India are characterised by inclement weather, inhospitable terrain, etc. In fact winter fogs, summer dust-storms, monsoon landslides only intensify with every passing year. The availability of indigenous navigational capability for positional and navigational information on highways, railways and airways as well as for disaster mitigation and a host of other civilian requirements in addition to military needs would dictate the need for capabilities designed to our typical needs and requirements. The availability of an indigenous navigational system covering at least the north-west and north-east of India is imperative for both military and civil requirements.

Financial propriety in investing in costly indigenous navigational capabilities would prima facie appear questionable. However, it needs to be borne in mind that apart from its own satellite-based augmentation system GAGAN (GPS and Geo-Augmented Navigation) which is strictly devoted to civilian uses only, India has already invested in every existing navigational system ranging from the American NAVSTAR Global Positioning System (GPS), Russian GLONASS to European GALILEO with no reciprocal guarantee of assured access. Thus instead of ignoring its own capabilities and investing in multiple agencies, it would certainly augur well to invest in national navigational capabilities as a long-term solution. While the requirements and costs of an indigenous navigational capability are debated, the following interim measures could be undertaken.

- Acquire navigation receiver systems compatible to multiple transmitters since receivers would need to be compatible with American, European systems, etc.

- Take advantage of existing GPS using differential GPS for greater reliability.

- Develop a system analogous to the Wide Area Augmentation System (WAAS)o of the USA for military uses possible. The system is a network of precisely surveyed ground reference stations which receives GPS signals and determine errors and compute corrections. These corrections are then transmitted from a geo-stationary communications satellite on the same frequency as GPS. This would enable continuous use of GPS even in the eventuality of the service provider attempting to deny or degrade GPS facilities. Incidentally, until the maturing of their indigenous space-based navigational system 'Beidou', the Chinese are known to be following the above option.[2]

- Install GPS integrity monitors at Air Traffic Control centres, etc. to monitor GPS, and enable corrective action in case of systems degradation or denial.

- While the above would be sufficient as temporary measures, in the near term a limited nav-sat constellation akin to the Japanese QZSS or the Chinese Beidou would need to be made available.

Early Warning

The requirement of EW against BM launches would require an enormous investment since neither would the satellites be dual-use nor would the requirement suffice by a single satellite, two satellites or part constellations. An enormous effort and investment into a constellation covering the entire subcontinent including its island territories would be required since a successful BM attack would have the same effect politically irrelevant of whether the target were Delhi or Deolali or Dibrugarh or any other obscure place. Such an enormous investment in the near-term is not considered prudent and it would be wiser at present to invest in capabilities like communications, navigation, observation, etc. which are of a more emergent and rewarding nature in the near term and the foreseeable future. Dispensing with an elaborate BMEW system in the near term is being suggested primarily on the following grounds:

- BMEW systems are largely a subset of BMD which itself is a controversial endeavour. The technology is yet to mature, the operational utility and concepts are unproven whereas initial and recurring costs enormous. Once the experimentation stage ends, depending on the failure or success of the endeavour, efforts may be put in.

- Unlike in the case of Cold War, BM rivals like the USSR who were separated by intercontinental distances and needed an elaborate investment in BMEW systems, our threat is characterised by proximate, contiguous neighbours. The prevailing conventional radars are capable of detecting and tracking missile trajectories within the atmosphere and whether such an enormous investment needs to be undertaken for tracking and detection beyond the atmosphere is a moot issue.

- Apart from BMs as a delivery option, numerous other options are available to both the BM and nuclear powers in our neighbourhood.

- In the near term, feasibility and applicability of BMEW may be studied in detail and towards mid-term, decision on the matter may be taken.

Protecting and Securing Assets in Space

As of now, no known protection measures have been undertaken to secure our assets in space. A hard kill (e.g. ASAT attack) or soft kill (jamming, interference) by a hostile entity would cause an insignificant dent in our military capabilities but an enormous dent in our economic capabilities. Increased use and exploitation of space capabilities for economic and other development would call for enhanced protection of space assets and capabilities; hence steps to protect and secure assets in space need to be undertaken in the near-term itself.

Threats Envisaged

Presently, threats to space-based systems are perceived in the following areas;

- Ground, airborne and space-based energy weapons like DEW, EMP, HPM,[2] etc.
- Kinetic kill by non-nuclear ASATs and other interceptors.
- Degradation by jamming and other forms of interference.

Effective threat levels of the above three would, however, vary. Technology in the first case is yet to mature completely and is extremely costly. The other two are more problematic due to prevalence of established technologies, which are capable of jamming or degrading space systems by targeting signals, uplink/downlink, etc. or actually destroy satellites. It also needs to be borne in mind that latent ASAT capabilities already exist with many space powers, conflicts and disputes are already in vogue as also is hijacking of signals, etc. The situation is already alarming and is only likely to worsen further.

Protection and Security of Space Assets

This would demand parallel defensive measures which are a studied, deliberate, institutionalised effort rather than just knee-jerk response to adverse situations. Our space systems would need to be equipped with proper shielding, frequency agility, manoeuvrabil'·y and encryption to be invulnerable and more importantly, survivab.e even with degraded capability in a worst-case scenario. Protection of space systems would largely revolve around the following:

- Orbital monitoring and protection
- Link (uplink/downlink) control and protection
- Allied receiver system (terrestrial) protection.

Protection Measures

Broadly, the following measures would need to be undertaken to ensure protection of space assets:

- Monitoring of space for continuous information on the location of satellites, space debris, asteroids and other harmful matter. This would necessitate the formation of aerospace surveillance centres which would surveill the entire vertical expanse inclusive of the atmosphere, near-Earth orbit and even beyond.

- Incorporation of survivability measures for both satellites and their payloads like hardening, shielding, etc. to shield against soft-kill energy weaponry like DEW, Lasers, HPM. ECM measures like frequency hopping, antenna nulling, etc. would need to be incorporated. Ability to pinpoint and enable counter-response against jamming, interference, etc. would also need to be factored into.

- Incorporation of additional energy (fuel) to provide manoeuvrability to defeat physical attack (hard-kill). Payload penalties would have to be designed within acceptable limits so as to compensate for hardening, manoeuvrability, etc.

- Encryption of satellite links (uplink/downlink) to protect and prevent intelligence leaks from transponders, transmitters, receivers, etc.

- Multi-sensor data fusionability to provide for redundancy and integrate satellite imagery and other data derived from airborne sensors like aircrafts, UAVs, etc. to defeat camouflage, concealment, deception and also compensate for degradation.

- While a wide variety of survivability and protection options exist, a judicious mix of efforts in response to the nature and probability of threats as well as the value of satellite would need to be undertaken for balancing conflicting requirements of security and cost. Towards mid-term, the technological maturing of micro and nano-satellites would herald a revolution in the uses and cost of satellites and consequent cheaper options would need to be explored.

Challenges to Civil–Military Modification

In the near term, the main differences perceived for modification of our prevailing space programme to meet security requirements relate to:

- Greater robustness and manoeuvrability to secure space assets against degradation, disruption and destruction by enemy counter measures like jamming, ASAT weaponry, etc.

- Increased resolution capabilities coupled with more frequent revisit capabilities to meet intelligence, targeting and other requirements.

- Independent, secure, dedicated and redundant communication and navigation links to ensure uninterrupted access even during times of crisis and wars unlike in the case of the prevailing US GPS, European Galileo, etc. whose use may be denied, restricted or even degraded by the service provider itself.

- Sharing of technological, operational and related know-how for building aerospace surveillance capabilities.

Organisational and Infrastructural Requirements

Apart from the establishment of a centralised coordinating agency like an aerospace command on which the consumption of the above goals would rest, certain other requirements would also have to be met, such as:

- Development of comprehensive defence policy and strategy on space. This would enable continuous review and optimal use of space assets.

- Development of doctrines and concepts driving correct and optimal use of space by the military.

- Development and training of human resources pool for optimal management and conduct of space operations.

- Development and shaping of infrastructure to fulfil the above.

Mid-Term (2017–2027)

It would be too ambitious to expect total fulfilment of all near-term goals towards the end of the year 2017, and it would be prudent to expect a spill-over of incomplete programmes on to the subsequent phase. The pace of technological and geopolitical change would dictate intense inherent dynamism and adaptability in our plans and capabilities. Nevertheless, by mid-term, it would be fair enough to expect a certain level of mission fulfilment in roles of communication, navigation, ocean reconnaissance, etc. A certain level of maturity in the technology, operational concepts, human resources, organisation and infrastructure may also be expected which would be instrumental in transferring and translating space capabilities into terrestrial military power.

Hence, by mid-term, the emphasis would shift to completion of pending projects and better integration of space onto conventional military capabilities. This would dictate a review of changed capabilities and focus on doctrine, strategies and procedures for optimal exercise of military power in pursuit of national objectives.

By mid-term, the focus would be on integrating not only the capabilities but also the strengths of space and conventional military capabilities which would enable responsive delivery of the cumulative potential of our conventional military and space power. The above integration would offer a military apparatus comprising

- Space-based systems aimed at force enhancement.
- Intermediate systems inclusive of:
  - Multispectral data relay system for data transmission amongst satellites, airborne and terrestrial sensors, weapon platforms and other support systems.
  - Multi-sensor data fusion capability.
  - Efficient data processing, storage and distribution systems.
  - A responsive and dynamic C4ISR system that enables rugged and redundant linkages between space and terrestrial systems.
- Conventional military systems inclusive of:
  - Components of air power.
  - Components of land power.
  - Components of sea power.

The above military apparatus is conceived to be dynamic and multilayered with ground assets, aircrafts, UAVs, other flying platforms and satellite networks. The structure would be designed for redundancy amongst a variety of platforms, overlapping coverage amongst a variety of sensors and connectivity by common C4ISR architecture. Such a structure would fulfil the requirements of detection, battlefield situational awareness, precise discrimination and selection of targets, precise munitions and platform delivery, near accurate BDA, etc.

Long-Term (2027–2037)

Long-term predictions and visions are fraught with difficulties on account of the pace of change. Extended long-term visions generally dissipate into mirages and hence for the present it would suffice to have a broad long-term concept which would deal largely with going beyond incremental capabilities and developing new strategies based upon capabilities foreseen in the future. Such an endeavour could be undertaken towards the end of the middle term and hence is dispensed with in the present.

Notes and References

1. John Pike, "The Military Uses of Outer Space", SIPRI Year Book 2002, Ch. 11, p. 636.

2. Directed Energy Weaponry, Electro Magnetic Pulse, High Power Microwave, etc.

Chapter 9
Leaders for Managing Future Space Capabilities

It is reasonably well known that space-enabled information and information-enabled warfare is the crux of modern military war fighting, the much-touted Revolution in Military Affairs (RMA) of the 1980s, Net Centric Warfare (NCW) of the 1990s and finally Defence Force Transformation Architecture concepts of the new millennia. Briefly, the very base of most of these concepts is contingent to the exploitation of space. Talk on the above concepts has been rampant in armed forces across the world (and ours) for quite some time. As a matter of fact, a veritable cottage industry has grown around these concepts. However, the pace of translation of concepts into military operations has been significant only in case of a few countries like the US and the FSU.[1] Most countries including ours are yet to venture beyond healthy discussion and debate into actual development of capabilities. In most cases this is primarily because of the usual inhibiting riders of high cost and technology as well as immature doctrines and inadequate comprehension of the operational utilities afford by integration of space into military capabilities.

Nevertheless, the impact of these riders in our unique context is not as significant considering that extant space capabilities do exist and a growing economy would be able to sustain reasonable military demands. Matching US levels of space utilisation to revolutionise military operations apparently is too ambitious, but making a beginning certainly is not. This assumes greater significance considering that a more intense, more powerful and yet less famous revolution in civilian affairs is already under way. The revolution is driven by our civilian space capabilities and is silently revolutionising national well being, commerce and development as never before. However, the development of national space capabilities and their integration into civil development is apparently not concurrently matched in a military context. Far from concurrent matching, the reality is that there exists a serious mismatch.

One of the primary factors contributing to the mismatch is the absence of leadership, both organisational and institutional for integrating space into military operations. This has consequently led to absence of an institutional space vision and absence of resources (both human and otherwise) for fulfilling aspirations. Fulfilling this lacunae would be essential for systematically integrating space into conventional military capabilities and then refining the broad strategy and operational doctrines for translating concepts of RMA, Net Centric Warfare (NCW), etc. into operational solutions.

Challenges to Leadership

While fulfilment of the leadership vacuum appears a fairly simplistic solution in theory, actual practice would be fraught with enormous difficulties and challenges. Inclusion of the space paradigm into military affairs and the near-chaotic pace of change and confusion in generates would present enormous leadership challenges at a variety of levels ranging from national organisational to institutional levels of squadrons, units, personnel, etc. The inherent pace of technological innovation in the realm of air and space is so intense that even aerospace super powers like the US and the FSU are yet to fully come to terms with the immense potential afforded by the vertical dimension, the challenge in integrating the mediums of air and space, the doctrinal challenges, the challenges to operational integration and a host of other issues.[2]

The complexities of integration and utilising space for military purposes assume greater significance in our context considering the fact that unlike in case of most countries wherein space programmes evolved to serve military needs, our space programmes were tailor-made purely to support civilian rather than military endeavours. Hence, civil to military spill-overs are by exception rather than design. Redesigning for military purposes would neither be possible nor desirable or even acceptable. Our requirements in the present and near foreseeable future would demand optimal exploitation of extant capabilities and gradual progression of capabilities in keeping with the technological, cost and other limitations.

Hence, integrating and operationalising space as well as developing military-specific space technology in the future would demand a re-examination of the operational attributes afforded as well as the roles, missions and employability of space in support of national security objectives. Enormous change would be forthcoming; the changes would need to be comprehended and engaged in their entirety. The change driven by the advent of space would need to be viewed not in isolation but in relation to the dynamics of the prevailing geo-economic, geo-strategic issues, its impact on national empowerment in a broader context and on military instruments of power in a narrower context. Managing the change would call for effective leadership. The challenge to leadership would be in anticipating change, accepting change; keeping pace with change and finally in managing change to support objectives. In order to undertake the above, it would be essential to briefly dwell on optimally utilising existing capabilities, envisaging future capabilities and then explore the options for developing leaders to manage future space capabilities.

Optimally Exploiting Capabilities

At the outset, it first needs to be borne in mind that space-based assets as in case of aircrafts are complex, costly and scarce instruments which have multiple applications and also apply at multiple levels of strategy, operations and tactics, as well as during peace, crises and wars. Hence, a broad-based comprehension of the operational possibilities as also the technological limitations would be an essential pre-requisite for managing future space capabilities. This would dictate a deeply cooperative and even integrated approach between the military and the Department of Space (DoS) to understand the impact of space on military operations, as well as for building and managing capabilities. Space shares a large degree of commonality with airpower in terms of characteristics, attributes, doctrinal utility, etc. and hence airpower expertise would be an essential prerequisite to develop leaders responsible for managing future space capabilities.[3] A reservoir of airpower expertise is already available with the IAF, the same would need to draw upon the space expertise available in the DoS and an initial effort could be undertaken which would be primarily aimed at:

- Comprehensively exploring the realm of possibilities afforded by space.
- Based on the possibilities, enunciating a dynamic vision and a road map.

● Developing leaders to translate possibilities, vision and roadmaps into operational solutions.

● Application and integration of space into conventional capabilities for providing solutions to national security challenges.

This is a vast field and initially would primarily aim at fulfilling two aspects. The first would be generic space familiarisation and dissemination of operational utility and knowledge across the military, and the second would deal with detailed studies aimed at building capabilities. The latter would rely heavily on consultations and discussions with a large variety of experts in the DOS as well as in different components of our security establishments.

Future Uses Envisaged

We have a fairly advanced space programme for economic and development purposes which already provides inputs and resources for defence, especially in communications and imagery. Further programmes with dual use applicability are on the anvil. Space capabilities for defence, would largely be built around these endeavours. The challenge to leaders would be in optimally managing extant capabilities and in judiciously building future capabilities. Our utility in the near term and foreseeable future would largely be limited to missions of force enhancement. Broadly, the near foreseeable future would demand the following space capabilities.

Observation/ISR

● The continuously expanding reach of the Air Force would demand accurate targeting, intelligence and information for efficient delivery of military power. Our long-range precision strike capability would be only as good as the accuracy and timeliness of targeting inputs available. The expanding reach of the IAF also implies a vastly expanded number of targets from which to select priorities in keeping with operational goals. A larger geographical coverage would expand the number and type of targets thereby demanding precise Intelligence Surveillance Reconnaissance (ISR) capabilities which could only be augmented and made more efficient by space based assets. Political air space restrictions would restrict acquisition of targeting intelligence and hence to match strategic reach with adequate intelligence and targeting information, the availability of space-based assets would be imperative. Permanent positional Space-based IMINT and ELINT would complement and compensate for the prevailing airborne capabilities, which are transitory and temporary in nature. For example, while the entire spectrum of aerial platforms ranging from aircrafts, UAVs, Aerostats as well as balloons in near space are of a transitory nature, satellites would enable safe, permanent presence and persistent over-watch over area of operations during peace, crises and wartime. In brief the prevailing strategic breadth, reach and vertical depth of the IAF would be more efficiently exploited for national security and defence goals.

Navigation

- In view of the enormously expanding reach of the IAF, these capabilities would be necessary for enabling precise navigation, targeting and delivery of scarce and costly platforms, munitions, personnel, humanitarian assistance, etc. well beyond national borders thereby increasing the contingency support as well as contingency management options of the Government in keeping with our rise as a global power.

- Increasing crowding of the aerospace continuum would demand augmentation, assistance, etc. by navsats for aerospace management (in both civil and military terms) and eventual phasing out of radar-based manual systems.

- It would enhance the efficiency of special operations by enabling insertion of personnel and materials well beyond our borders as well as search and rescue, recovery of stranded combatants from deep within enemy territory.

Communication

- India's stated nuclear doctrine would demand availability of survivable communication links for a retaliatory strike by elements of the nuclear triad that could be provided only by space-based systems.

- The enormous strategic reach and mobility of IAF enables rapid response and insertion of military force at long distances and the same would demand extensive coordination, command and control that could be addressed only by space-based communications.

- Enormous information would be available from aerospace elements inclusive of aerial platforms like aircrafts, UAVs, satellites, etc. This would require real time, transfer and distribution vertically and laterally. Efficient, secure real time data transfer would dictate the need for communication satellites.

- Apart from enhancing efficiency of C4I, the Information-Decision-Action cycle, etc. the requirement of aerospace management would also demand real-time, immediate transfer of data that could be enabled by communication satellites.

Other Applications

The aforementioned are some of the operational gains forthcoming on harnessing of capabilities of air and space power. Besides, numerous other applications are available though untapped. Space applications as previously mentioned have multiple applications and would be of use across a wide spectrum of occupations, branches and trades. For example, prevailing telemedicine applications would need to be exploited by the medical branches (as well as the general populace), education officers could tap the immense potential of tele-education, signals officers could utilise the immense potential of satellite communications for an independent mobile communications systems,[4] administrative officers could use GIS systems for estate planning, formulating Key Location Plans (KLP), etc. flight safety officers could use GIS applications for plotting power grid lines, helipad location and a host of other utilities. Junior-level leaders like Unit Warrant Officers (UWOs) would have immense use of GPS locator chips to keep track of their work force, Senior Non-Commissioned Officers (SNCOs) in charge of mechanical transport, catering, etc. would have immense uses of space-based location and positioning devices to keep track, manage and distribute resources more efficiently. The overall impact on efficiency would be enormous. The best manner to promote integration would be by demystifying space, making space capabilities familiar and by broadening the space user base, no matter how rudimentary the uses and applications. Increasing use in multi-fold aspects would increase the employability and utility awareness across the IAF and the armed forces thereby enabling systematic integration at a later stage. Once space capabilities become demonstratively apparent and make their presence felt, the demands for greater and better integration would automatically follow. These might initially result in chaotic growth and demands for services. The challenge to leaders would be to anticipate the changes and in judiciously balance conflicting requirements of demand and supply.

Translating Space Capabilities into Strengths

Comprehensive exploitation of space and concurrent building of capabilities and their conversion into strengths would demand intense broad-based efforts in terms of developing human resources, training and education, familiarisation of missions, concepts, uses and a host of other factors. The challenge to leadership would be in terms of building, sustaining and managing space capabilities and enabling percolation of these capabilities to the lowest level possible. Penetration of space-based facilities into cockpits, soldier back-packs, operation rooms, etc. would demand enormous efforts and training to disseminate operational know-how and widen user participation and involvement. Increased user involvement would ensure greater demonstration of capabilities leading to greater demands, greater integration and greater operationalisation of capabilities. Nevertheless, ad hoc and haphazard growth would be self-defeating, containing the conflicting demands of widening user participation and streamlining the supply and demand chain would be essential to avoid chaos and confusion. The test of effective leadership would be in managing conflicting demands, balancing requirements and ensuring judicious procurement and distribution.

Considering that the above venture would continually be impacted by new and unfamiliar technologies, enormous efforts would have to initially go into basic familiarisation, adaptation and broad dissemination of missions possible by integrating new technologies, evolving concepts, etc. This effort would have to ensure that the dynamics of technological change and evolving operational requirements are factored into at planning stages itself for lesser wastages on consummation of capabilities. Foresight, vision and ingenuity tempered by adequate knowledge and analysis would distinguish leaders tasked with managing future space capabilities.

Developing Leaders for Managing Future Capabilities

To translate capabilities into operational strengths, purposeful and deliberate efforts would have to be put in to develop leaders who are adept at converting concepts into operational solutions, are knowledgeable and possess cross-competencies in a variety of disciplines. These leaders would gradually comprehend the desired mix of space power competencies and provide space solutions to operational problems. Demands presently and in the future would exceed supply in case of air and space assets as well as personnel, hence development of cross-competencies and multi-tasking would be the norm rather than the exception.

Correct identification, education, training and formation of a human resources pool would be the key to developing space competencies. In essence the foundations for human resources development would rest with the IAF on one hand and the DoS on the other hand. The requirements could be centrally collated by the IAF and after due deliberation could be developed upon in consultation with the DoS. Such an endeavour would fulfil the requirement of an educated and trained leadership pool which would be charged with translation of possibilities into capabilities.

Building the Leadership Pool

Akin to air and space assets, military human resources are a scarce commodity. The leadership pool tasked with converting space possibilities into capabilities would enhance need to be sourced or multi-tasked (at least initially) from prevailing resources and thus the model in Figure 9.1 is proposed.

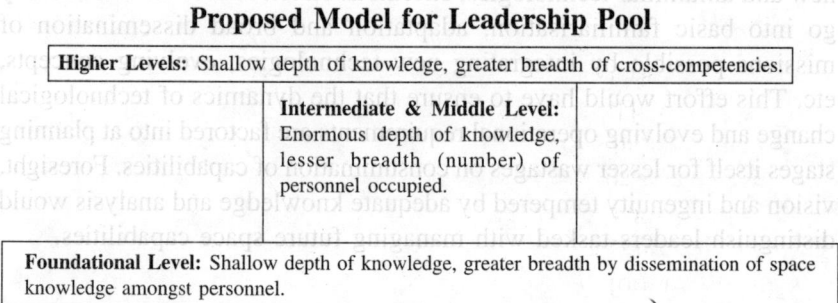

Proposed Model for Leadership Pool

Higher Levels: Shallow depth of knowledge, greater breadth of cross-competencies.
Intermediate & Middle Level: Enormous depth of knowledge, lesser breadth (number) of personnel occupied.
Foundational Level: Shallow depth of knowledge, greater breadth by dissemination of space knowledge amongst personnel.

Figure 9.1

Foundational Level

The emphasis here would be on familiarising space to a wide cross-section of personnel and building a reservoir of people aware of the impact and utility of space to military operations. The idea would be to demystify and democratise space capabilities. This could be fulfilled by a two-pronged approach. On the one side basic space familiarisation courses could be conducted at training and educational institutes and on the other hand the same could be incorporated at induction stage itself. For example, while elementary space capsules could be introduced along with General Service Knowledge (GSK) and basic airpower syllabus at induction stage itself to familiarise coming generations, prevailing generations could be educated by conducting the same at CAPS, CAW, or any other suitable institute. The course content could be suitably modified to apply both to officers and enlisted personnel. The depth of knowledge would be shallow but breadth of space familiarisation would be enormous, consequently the pool of resources to draw upon for building space capabilities would be enormous over a period of time. Rotation and integration of space qualified leaders would lead to a domino effect thereby increasing the resource base. Later space specialists of the middle and intermediate level would be extracted from this general pool for managing space capabilities. Over a period of time the IAF could become the lead service in providing quality space education (in terms of application to security needs) not only within the IAF or the military but also to other security agencies like the paramilitary forces, police, intelligence agencies, etc. After all, space applications transcend a variety of applications ranging from military specific strategic imagery to even crop analysis to detect illegal cocaine/opium plantations hidden in farms, tracking prisoners, jail parolees, emergency services, disaster management, and so on. This endeavour would thus provide the foundation upon which space competencies for security applications are built over a period of time.

Middle and Intermediate Level

As the name suggests this would target leadership at the aforesaid levels. The emphasis at this level would be on developing leaders with an expertise in particular niche areas related to space like particular mission areas of navigation, communication, ISR, etc. as well as areas related to concepts and doctrines, technology, space law, operational integration and application, planning, training and education and policy. The endeavour would be to develop a narrow core of specialists with enormous depth of knowledge and expertise in their respective fields. This in the near term would be the most challenging arena considering that there exists no specialisation and competencies would have to be built upon as collaborative efforts with the DoS and other foreign training institutes. Notwithstanding the same, within a period of around five years or so it would be possible to identify, select, train and develop leaders with expertise in niche areas. These leaders would form the core of space specialists who would be tasked with integrating space and developing space competencies. These leaders would be responsible for vetting the possibilities, mitigating the technological challenges, adapting technology and doctrine, streamlining demands and acquisition, formulating policies and tackling a host of other issues which would demand expertise, innovation and ingenuity. This level would over a period of time provide a ready leadership cadre for enabling vertical and horizontal integration of space into main-stream military operations. The challenges related to career progression, personnel matters and growth would have to be suitably met considering that this core would need to be characterised by continuity and specialisation.

Higher Levels

The emphasis here would be on developing leaders with a reasonably lesser depth of expertise but enormous breadth of awareness regarding space related affairs. This could be obtained by providing requisite space education as well as brief space assignments to develop leaders with cross-competencies in as many space disciplines as possible. Brief assignments would not compensate for expertise, but a balanced mix of education and exposure would develop leaders with the requisite competencies to grasp the contextual elements of issues, problems, etc. While initial efforts would be focussed largely on education, as capabilities develop, a judicious mix of education and exposure could be undertaken. Over a period of time, these levels would also be fed by competent experts from the intermediate and middle level of space specialists.

Familiarising Space

It would be imperative to familiarise leaders with space capabilities and increase military space presence as much as possible to ensure optimal comprehension, application and exploitation of capabilities. It could primarily be undertaken by:

- Including space in education and training by designing suitable courses and curriculum.
- Exposure to space capabilities, possibilities and applicability. This would entail cross-training and study tours at ISRO and other institutes of the DoS.
- Incorporating space capabilities in exercises, war games, simulations, etc.

Education, Training and Integration

There exist numerous definitions and interpretations of education and training. However, for our present purpose, the premise that education serves to provide knowledge and a broad intellectual understanding of issues whereas training provides the requisite skills to apply knowledge would suffice. Both education and training would be essential prerequisites to meet the challenge of developing leaders to manage and integrate space capabilities. Optimal exploitation of prevailing capabilities and systematic building of capabilities would be dependent on the quality of these two factors. Education and training would not only refine the tactics, techniques and procedures for optimal exploitation, they would encourage use and employment of space capabilities to meet operational challenges. Education would enable comprehensive understanding of the impact of space on military operations and proper training would ensure gradual integration, incorporation of space capabilities into both operational combat as well as support units.

Overall, it would enable the development of leaders who use their specific branch/trade skills to collectively produce and disseminate space-based information and other facilities for integration into routine day-to-day tasks and specific operations. For example, personnel devoted to imagery processing, navigation, etc. would provide information for combat military operations whereas personnel in other areas could use tele-medicine, tele-education, etc. for general utility of personnel and their families. The impact on military efficiency would be manifold.

Space education would demand a judicious mix of operational utility and technical education. Personnel would need to be educated by means of preliminary, intermediate as well as advanced education. The emphasis of education would need to be on operationalising and integrating space rather than space sciences. As a consequence, the accent of education would be more on how space enhances military efficiency, on its utility and less on orbital mechanics, systems engineering, etc. A rudimentary knowledge of space science to enable comprehension of the technological limitations and operational possibilities would suffice. In our unique case, the entire space programme is managed by the DoS and hence our focus (initially at least) would be on utilising extant capabilities and judicious planning for future capabilities unlike in case of countries like the US or Russia wherein the space programmes are actually run by military bodies and hence demand enormous efforts in space technology education. The point is operator-level education rather than launch or systems designer level would suffice. The above is validated considering the fact that even hugely successful and highly complicated military space systems like NAVSTAR GPS are operated by personnel without engineering or even science degrees. Most satellite vehicle operators do not hold (and are neither required to hold) technical degrees. In fact, satellite systems operators who generate and transmit commands to satellites are young airmen, often only a few months out of basic training and possessing only high school diplomas.[5] By contrast, the IAF already has a vast pool of technologically qualified engineers and personnel who could adapt their competencies to include space operations. In view of the foregoing, the following education model depicted in Figure 9.2 is proposed for incorporation at various levels:

Proposed Education Model

Levels	Yrs of Service	Officers	Airmen	Hours	Content
Induction	00	Pre-commission	Pre-enlistment	10	Basic
Entry	1–2	Branch Specialisation	OJT/equivalent	20	Intermediate
Junior Leaders	2–6	Junior Commanders	Equivalent	20	Intermediate
Middle level	6–13	Customised Space Course	Customised Space Course	30/30+	Advanced
Senior Leaders	13+	Customised Space Course	Customised Space Course	20	Intermediate
Air Ranks	—	Space Familiarisation	—	10	Broad-based

Figure 9.2

This model is conceptual rather than instructive and based on the concept further refinements could be undertaken. A suitable space curriculum could be evolved for the purpose.

Apart from education, personnel would need to be trained in a variety of areas to provide skills required to plan and conduct space operations. The primary emphasis would again be on integrating space-based information capabilities, hence skills would need to be honed accordingly. Practising skills aimed at providing space-based information, analysing capabilities and limitations, recommending space applications to support operational requirements and decision making, etc. would provide the decisive edge in conflicts.

Simulated Training and War Gaming

However, education and training by themselves would be artificial experiences and affordable virtual (or real-time) experiences with effective space models and simulation would need to be integrated into field level exercises, war gaming, etc. The models and simulation would need to be as realistic as possible and should attempt to realistically duplicate extant space capabilities. This would enable near-accurate comprehension of impact and incorporation of space capabilities into operations. Based on this, leaders could plan for, demand, distribute and use capabilities accordingly. To begin with, space visualisation and analysis kits like the Satellite Tool Kit (STK), etc. could be procured off the shelf.[6] Later on software-based on our own capabilities and environment could be designed for our unique purposes. This would enable reasonably accurate modelling of our space systems, their capabilities, vulnerabilities and impact on military efficacy and national security. During major exercises, actually available capabilities could also be tasked for assessing the efficacy of systems, their navigation into military capabilities and their overall impact.

Application of Knowledge

Technology is a double-edged sword and serves those on the lower as well as higher end of the spectrum. The factor distinguishing its exploitation is the extent of knowledge and application of knowledge to obtain requisite solutions. Integrating space education and training across the IAF would also ensure that personnel apply their space knowledge to operational tasks and subsequently come up with ingenious and innovative applications. Numerous instances of innovative applications of knowledge to security needs exist. For example, the impact of Indian ingenuity was demonstratively manifest in May 1998 when India surprised the world with its nuclear tests in spite of persistent coverage by US spy satellites. American space power had clearly failed to deter India from conducting nuclear tests. This was primarily because the implications of satellite overpass were known and understood well. Operational application of knowledge enabled India to trick US satellites and conceal the tests from US satellites by conducting nuclear tests "when sand storms normally swept across the Thar Desert and intense heat could disrupt surveillance sensors. Activity was also timed around the flights of spy satellites".[7] Similarly, while the impact of space systems like GPS and Precision Guided Munitions (PGMs) like HARM air-to-ground missile is well known, what is less famously known is that innovative use of domestic microwave ovens, low power GPS jammers, etc. were sufficient to blunt the impact of high technology space systems. During the allied forces Yugoslav campaign, based on Russian advice, the Serbs lured away NATO aircraft and PGMs from their intended targets by switching on microwave ovens and aiming them upwards. American HARM missiles would home in on any strong source of radio emission in the 400-10,000 MHz range, exactly the range of conventional household microwave ovens. Hence, the Serbs used household microwave ovens to simulate the emissions of armoured transport systems and lured away NATO planes from their intended targets domestic microwaves.[5] The point is that Russian knowledge of space systems enabled blunting of the American space offensive.

Gains Envisaged

Correct identification, education, training and formation of a human resources pool would be the key to developing space competencies over a period of time. Such an endeavour would broadly fulfil the following requirements:

- Streamline the selection, education and training for developing leaders who would form the core group committed to developing space concepts and applying them to resolve national security issues.

- Increase the understanding of the impact of space on military operations by education, training, battlefield simulation, participation in exercises, etc. This would serve to familiarise space operations across the rank and file. With adequate encouragement it would enable a reciprocal flow of ideas from across the Air Force and other agencies which could be developed into operational concepts.

- Develop leaders who understand the changing dynamics of national security issues, adapt space competencies to support national objectives and plan development of space capabilities accordingly.

- Develop leaders who envision, develop, manage, acquire, sustain, support and employ capabilities which exploit the space domain to create military effects.

- Develop leaders to implement space power doctrine and strategy.

- Develop leaders to strengthen space capabilities for credible nuclear deterrence.

- Enable systematic and professional development aimed not only at producing space leaders, but also to provide the foundation upon which aerospace competencies are built over a period of time.

Conclusion

The foregoing recommendations are by no means exhaustive and would primarily serve to broadly guide the development of space competencies at the prevailing nascent level. Further refinements and incremental development of capabilities would follow once the leadership issue is deliberated upon and resolved. Resolving the leadership issue would be the first step to building aerospace competencies and heralding any form of a revolution in military affairs or networking.

Notes and References

1. The Former Soviet Union (FSU) is not known to have evolved beyond the RMA stage and as of 2004, its space capabilities and overall military capabilities have seriously dwindled.

2. The complexity of the issues can be gauged considering the fact that the world's leading military space driver; the US Air Force within a span of 10 years (1990–2000) attempted thrice to integrate space into an air and space

force and yet did not meet unqualified success in their endeavours. See Lt Col Mark P. Jelonek, USAF, "Toward an Air and Space Force", *CADRE Paper* AU Press, Maxwell Air Force Base, Alabama.

3. For a detailed brief on the commonality between air and space, see Sqn Ldr KK Nair, "Merging Frontiers of Air and Space", *Airpower Journal,* Vol. 2, No. 3 Monsoon 2005.

4. Instead of relying on commercial vendors like Airtel, Hutch, etc. for mobile connectivity, it would be worthwhile to explore the possibility of an independent IAF cell phone network. Considering that most stations have VSAT terminals and the frequency spectrum is available, a network could be developed with confidential connectivity for personnel and open channels for families. This would serve purposes of both security as well as welfare.

5. See 1st Lt Brent D. Ziainik, USAF, "Mahan on Space Education", *Air & Space Power Journal,* Winter 2005.

6. For details on computer-generated space software and models, see site of Air University, Center for Space Studies at http://space.au.af.mil/teaching.htm

7. See Krishnan Guruswamy, "India tricks US Satellites", *Associated Press,* 19 May 1998. Available at http://abcnews.go.com/sections/world/Daily news/india980519_nukes.html.

8. Vladimir Bogdanov, "Anti-weapon: Russian scientists threaten to halt space war", *Rossiyaskaya Gazeta* (Moscow), 18 October 2002 quoted by *Global Research.Ca* November 2002 at www.globalresearch.ca/articles/BOG211A.html